Urban and Industrial Water Conservation Methods

Urban and Industrial Water Conservation Methods

Abbas Yari, Saeid Eslamian, and Faezeh Eslamian

CRC Press
Taylor & Francis Group
Boca Raton London New York

CRC Press is an imprint of the
Taylor & Francis Group, an **informa** business

First edition published 2021
by CRC Press
6000 Broken Sound Parkway NW, Suite 300, Boca Raton, FL 33487-2742

and by CRC Press
2 Park Square, Milton Park, Abingdon, Oxon, OX14 4RN

CRC Press is an imprint of Taylor & Francis Group, LLC

Library of Congress Cataloging-in-Publication Data

Names: Yari, Abbas, editor. | Eslamian, Saeid, editor. | Eslamian, Faezeh A., editor.
Title: Urban and industrial water conservation methods / edited by Abbas Yari, Saeid Eslamian, Faezeh Eslamian.
Description: First edition. | Boca Raton : CRC Press, 2021. | Includes bibliographical references and index.
Identifiers: LCCN 2020020221 (print) | LCCN 2020020222 (ebook) | ISBN 9780367533182 (hbk) | ISBN 9781003081531 (ebk)
Subjects: LCSH: Water conservation.
Classification: LCC TD388 .U73 2021 (print) | LCC TD388 (ebook) | DDC 628.1/3--dc23
LC record available at https://lccn.loc.gov/2020020221
LC ebook record available at https://lccn.loc.gov/2020020222

ISBN: 9780367533182 (hbk)
ISBN: 9781003081531 (ebk)

Typeset in Times
by Deanta Global Publishing Services, Chennai, India

Contents

Preface

Replacing all standard residential appurtenances with water-efficient products would result in an overall decrease of 32% in domestic water consumption, or 40,716 liters per year for a European Union (EU) household, according to the EU Directorate-General for the Environment. What is the situation for other urban water consumers? How great are the conservation potentials for each sector? What are the results of case studies and launched projects, globally? How could the use of water conservation devices and methods make real changes possible in each sector? What are the global water efficiency regulations for residential buildings and what are their contents? In this book, all urban water consumers are discussed with respect to the extent of their water use, their usage characteristics, and their conservation devices and practices, in detail.

Access to dependable water resources is a major production factor for most economic sectors in the world. Residents, manufacturing plants, agriculture farms, and tourism rely on a reliable supply of water, often of a pre-determined quality. Since water scarcity and droughts are becoming issues of high priority for authorities, regardless of industry, water-saving is also becoming more important. Thus, there is an urgent call for administrations to address this problem to ensure a clear, sustainable future for water resources and management. Reliable and comprehensive data is a key for policymakers to adopt a successful strategy to conserve water.

A good understanding and proper use of water conservation devices and practices are essential to achieve the goals of a water conservation plan. The optimal and wise use of water resources is one of the many essential actions needed, especially for processed water that is often transported to urban and rural areas, mostly through pumping, which results in a high energy consumption. Indeed, many appliances are wasteful and inefficient.

Authorities' requests for advice on water conservation through the optimization of devices and practices led us to work theoretically and in practice on the different kinds of solutions. Consumers can choose from a wide range of technical solutions to save water with minimal changes to their systems or behavior. Our studies have shown that there is a significant potential in such devices and policies to reduce consumptions, which has encouraged us to write a technical book to comprehensively illustrate the advantages of these solutions.

Specific aspects of conservation and efficiency in water use in commercial, institutional, and industrial sectors are also discussed in this book. The conservation and efficiency of domestic water use in these sectors is similar to that of residential applications. Many companies can successfully operate on 20–50% of previous water consumption by minimizing water use. Cost-effective, water-saving devices and practices, some with payback periods of only a few days, are also discussed. The discussion highlights the typical water savings that can be achieved for commercial, institutional, and industrial applications and explains how to identify the most appropriate devices and practices for specific equipment, processes, or sites.

The authors would like to express a special appreciation to Dr. Alireza Sarabi, Dr. Mohammadreza Askari, Mr. Vahid Faridani, and Mr. Ali Pourfarzaneh in Bandab Consulting Engineers for their contributions and close collaboration in completing and presenting this book.

Abbas Yari
Bandab Consulting Engineers
Tehran, Iran

Saeid Eslamian
Center of Excellence in Risk Management and Natural Hazards
Isfahan University of Technology
Isfahan, Iran

Faezeh Eslamian
McGill University
Montreal, Canada

Authors

Abbas Yari is a project manager and head of both the engineering design team and the water supply department in Bandab Consulting Engineers. He has been involved in strategic national projects as the project manager and senior engineer. He is the co-editor of Iran National Standards for "Analysis and design of concrete water storage reservoirs" and has written several technical papers in local journals.

Yari has 30 years' professional experience in the fields of water supply for urban/rural, industrial, and agricultural development and water conservation in residential, commercial, institutional, and industrial sectors, as well as in rehabilitation projects including transmission pipelines, distribution networks, pumping stations, instrumentation, telemetry and control philosophy, reservoirs, tanks, large-scale pressurized irrigation networks, intakes, water-saving devices, reverse osmosis desalination plants and brine transmission and disposal, dam bottom outlets, leak detection, water hammer analysis, equipment selection and commissioning, etc.

Saeid Eslamian is a full professor of environmental hydrology and water resources engineering in the Department of Water Engineering at Isfahan University of Technology, where he has been since 1995. His research focuses mainly on statistical and environmental hydrology in a changing climate. In recent years, he has worked on modeling natural hazards, including floods, severe storms, wind, drought, pollution, water reuse, sustainable development and resiliency, etc. Formerly, he was a visiting professor at Princeton University, New Jersey, and at the University of ETH Zurich, Switzerland. On the research side, he started a research partnership in 2014 with McGill University, Canada, and, in 2019, with University Hawaii, USA. He has contributed to more than 600 publications in journals, books, and technical reports. He is the founder and chief editor of the *International Journal of Hydrology Science and Technology* (IJHST). Eslamian is now associate editor of four important publications: the *Journal of Hydrology* (Elsevier), *Eco-Hydrology and Hydrobiology* (Elsevier), the *Journal of Water Reuse and Desalination* (IWA), and the *Journal of the Saudi Society of Agricultural Sciences* (Elsevier). He is also the author of more than 40 books and 200 book chapters.

Eslamian's professional experience includes membership on editorial boards, and he is a reviewer of more than 100 *Web of Science* (ISI) journals, including the ASCE *Journal of Hydrologic Engineering*, the ASCE *Journal of Water Resources Planning and Management*, the ASCE *Journal of Irrigation and Drainage Engineering*, *Advances in Water Resources*, *Groundwater*, *Hydrological Processes*, the *Hydrological Sciences Journal*, *Global Planetary Changes*, *Water Resources*

Management, Water Science and Technology, Eco-Hydrology, the *Journal of American Water Resources Association*, and the *American Water Works Association Journal*. UNESCO nominated him for a special issue of the *Eco-Hydrology and Hydrobiology Journal* in 2015.

Eslamian was selected as an outstanding reviewer for the *Journal of Hydrologic Engineering* in 2009 and received the EWRI/ASCE Visiting International Fellowship in Rhode Island in 2010. He was also awarded outstanding prizes from the Iranian Hydraulics Association in 2005 and the Iranian Petroleum and Oil Industry in 2011. Eslamian has been chosen as a distinguished researcher in Isfahan University of Technology (IUT) and Isfahan Province in 2012 and 2014, respectively. In 2016, his work as a nationally distinguished researcher saw him put forward as a candidate for Iran's National Elites Foundation.

Eslamian has also been the referee for many international organizations and universities. Some examples include the U.S. Civilian Research and Development Foundation (USCRDF), the Swiss Network for International Studies, the Majesty Research Trust Fund of Sultan Qaboos University of Oman, the Royal Jordanian Geography Center College, and the Research Department of Swinburne University of Technology, Australia. He is also a member of the following associations: the American Society of Civil Engineers (ASCE), the International Association of Hydrologic Science (IAHS), the World Conservation Union (IUCN), the GC Network for Drylands Research and Development (NDRD), the International Association for Urban Climate (IAUC), the International Society for Agricultural Meteorology (ISAM), the Association of Water and Environment Modeling (AWEM), the International Association of Hydrological Sciences (IAHS), and the UK Drought National Center (UKDNC).

Eslamian finished Hakimsanaei High School in Isfahan in 1979. After the Islamic Revolution, he was admitted to IUT for a BS in Water Engineering, graduating in 1986. After graduation, he was offered a scholarship for a master's degree program at Tarbiat Modares University, Tehran. He finished his studies in hydrology and water resources engineering in 1989. In 1991, he was awarded a scholarship for a PhD in Civil Engineering at the University of New South Wales, Australia. His supervisor was Professor David H. Pilgrim, who encouraged him to work on "Regional Flood Frequency Analysis Using a New Region of Influence Approach." He earned a PhD in 1995 and returned to his home country and to IUT. In 2001, he was promoted to associate professor and in 2014 to full professor. For the past 25 years, he has been nominated for different positions at IUT, including university president consultant, faculty deputy of education, and head of department. Eslamian is now director for the Center of Excellence in Risk Management and Natural Hazards (RiMaNaH).

Eslamian made scientific visits to the United States, Switzerland, and Canada in 2006, 2008, and 2015, respectively. In the first, he was offered the position of visiting professor by Princeton University and worked jointly with Professor Eric F. Wood at the School of Engineering and Applied Sciences. The outcome was a contribution in hydrological and agricultural drought interaction knowledge by developing multivariate L-moments between soil moisture and low flows for northeastern U.S. streams.

Recently, Eslamian has held editorship of 11 handbooks published by Taylor & Francis (CRC Press): the three-volume *Handbook of Engineering Hydrology* in 2014,

Urban Water Reuse Handbook in 2016 (IWA partnership), *Underground Aqueducts Handbook* (2017), the three-volume *Handbook of Drought and Water Scarcity* (2017), *Constructed Wetlands: Hydraulic Design* (2020), *Urban and Industrial Water Conservation Methods* (2020) and *Handbook of Irrigation System Selection for Semi-Arid Regions* (2020). The two-volume *Handbook of Water Harvesting and Conservation* is a 2020 publication by Wiley that has been awarded the NYAS partnership. *An Evaluation of Groundwater Storage Potentials in a Semiarid Climate* and *Advances in Hydrogeochemistry Research*, both by Nova Science Publishers, are also among his book publications in 2019 and 2020.

Faezeh Eslamian holds a PhD in Bioresources Engineering from McGill University. Her research focuses on the development of a novel lime-based product to mitigate phosphorus loss from agricultural fields. Faezeh completed her bachelor and master's degrees in Civil and Environmental Engineering at Isfahan University of Technology, Iran, where she evaluated natural and low-cost absorbents for the removal of pollutants such as textile dyes and heavy metals. Furthermore, she has conducted research on worldwide water quality standards and wastewater reuse guidelines. Faezeh is an experienced multidisciplinary researcher with an interest in soil and water quality, environmental remediation, water reuse, and drought management.

1 An Introduction to Residential Water Users

Abbas Yari, Saeid Eslamian, and Faezeh Eslamian

Water withdrawals have tripled over the last 50 years due to rapid population growth. With water resources extensively strained or even depleted; however, proper management of supply and careful use of these resources has not received enough attention.

The measurement of consumers' water intake shows that about 65% of domestic consumption is due to the use of toilet siphons, showerheads, lavatory taps, and kitchen fixtures and appliances. Several studies around the world have shown that by installing water-restricting devices and limiting water flow, a significant reduction in water consumption can be achieved.

The intelligent selection of water conservation devices (WCDs) can play an essential and influential role in every water conservation project.

Showerheads and taps contribute 33% and 10% of average household water use, respectively. Replacing all standard residential appurtenances (faucets, toilets, showerheads, baths, washing machines, dishwashers, and outdoor water consumers) with water-efficient products would result in an overall 32% decrease in domestic water consumption, or 40,716 liters per year for a European household

By retrofitting Kashan's city fixtures, residential water consumption was reduced by 22%. As a result, Kashan could postpone its next water supply project by up to six years. The cost–benefit ratio of this conservation measure is estimated to be 5.8:1.

1.1 INTRODUCTION

Demographics and the increasing consumption that comes with rising per capita incomes are the most important drivers of pressure on water. The world's population is growing by about 80 million people per year, implying an increased freshwater demand of about 64 billion cubic meters per year. Competition for water exists at all levels and is forecast to increase alongside demands for water in almost all countries. With rapid population growth. This trend is explained largely by the rapid increase in irrigation development stimulated by food demand in the 1970s, and by the continued growth of agriculture-based economies. In 2030, 47% of the world's population will be living in areas of high water stress. Most population growth will occur in developing countries, mainly in regions that are already experiencing water stress and in areas with limited access to safe drinking water and adequate sanitation facilities (United Nations World Water Development 2009).

1.2 WATER DEMAND MANAGEMENT

Water demand trends in urban areas around the world show a continuous increase over the past few decades predicting continued growth in the coming years. The primary causes of increased water consumption and its supply problems might be population growth and urbanization, and consequently the expansion of cities, along with changes in lifestyle, demographic structure, excessive water extraction from all available resources regardless of their proximity, and also the possible impacts of climate change. It is challenging and sometimes impossible for existing resources to cope with the growing demand, especially in areas already faced with water stress or crises.

Though water resources are extensively strained or even depleted, proper management of the supply and use of these resources has not received enough attention. Obviously, water resources are limited and, as a rule, there is no suitable alternative for them. It is also a challenging task to find new sources or increase water extraction from current resources. To evade these severe crises, all authorities and citizens must be committed to rational and thoughtful solutions, and to changing their lifestyles towards optimal consumption.

Water consumption control can be approached in two different ways: the supply side, which is primarily the responsibility of the relevant public bodies, and the demand side, which mainly includes consumers.

Some of the supply-side management solutions are:

- Leakage and loss control in transmission lines and water distribution networks;
- Automation of pressure adjustments and leakage/burst management;
- Creating a smart network;
- Improving water use efficiency;
- Using reclaimed wastewater for irrigation and miscellaneous water users;
- Using surface water for urban landscapes.

For the development of new water resources, in addition to the limited and costly options, the sensitivity of citizens and policymakers to new projects (such as initiating or expanding withdrawal from existing water resources or interbasin water transfer) might be raised. Thus, demand-side management could be a more attractive option for the optimal use of existing water resources and should be put into practice. In this way, without imposing extra pressure on current resources, some parts of the available water resources and water supply systems can be released.

Water demand management involves a broad range of measures, policies, or investments for specific instruments to achieve the efficient distribution of water to all members of society. These actions could include:

- Water pricing reform;
- Promoting the use of water-reducing devices and motivating subscribers to install them;

- Improving water-consuming equipment;
- Mandating the installation of efficient water-consuming devices, mainly for new buildings, through legislation;
- Measuring the combined demand of consumers;
- Plumbing inspection programs that include the examination of all water-consuming appliances and the identification and repair of leaks in the buildings;
- Direct communication with consumers on an effecttive strategy, based on the socio-psychological recognition of the society and its values, including massive campaigns at the community level, public places, online, etc.;
- Providing consulting services and assigning incentives to citizens;
- Using treated water as an alternative to primary and potable water resources;
- Restrictions on water consumption for some water users, permanently or temporarily.

Demand-side management can be implemented in the short or long term, depending on the needs of the community. Limitations such as abnormal pricing for drought conditions and modernization programs with low-cost conservation devices are examples of short-term actions. Examples of long-term measures, among many other solutions, include modifying the pricing policy, identifying leaks, and repairing and adjusting the water efficiency of new buildings. Assessment of demand-side management practices is also critical to ensure that the measures are economical and cost-effective.

One of the most effective methods of monitoring water consumption is to focus on consumer water components. The measurement of consumers' water intake shows that about 65% of domestic consumption is due to the use of toilet siphons, showerheads, lavatory taps, and kitchen fixtures and appliances. Several studies around the world have shown that by installing water-restricting devices and limiting water flow through appliances, a significant reduction in water consumption can be achieved.

Controlling and reducing the use of bathroom showerheads, taps, and kitchen faucet with small and inexpensive devices can be easily achieved. In homes and places with a flush tank installed, the use of dual-flush siphons helps significantly to reduce water consumption. The intelligent selection of water conservation devices (WCDs) can play an essential and influential role in the success of research, studies, and action.

Regarding water efficiency and savings, the European Commission's study on water-efficient standards by the Directorate-General for the Environment estimated that showerheads and taps contribute 33% and 10% of average household water use, respectively. Replacing all standard residential appurtenances (faucets, toilets, showerheads, baths, washing machines, dishwashers, and outdoor water consumers) with water-efficient products would result in an overall 32% decrease in domestic water consumption, or 40,716 liters per year for a European Union (EU) household,

(Mudgal et al. 2009). Table 1.1 illustrates the technologies available to conserve water in urban water-using devices and recommends the best available and practical technologies, too (Savage 2009).

1.3 CASE STUDIES

Since the 1980s, Israel has been using drip irrigation and micro-sprinkler techniques to expand crop output within the limits of existing water supplies. These techniques are mainly used for vegetable and fruit trees, and are integrated into computerized systems that operate irrigation applications automatically based on information collected via plant moisture sensors. This technology, combined with the use of water-efficient crops, has resulted in an irrigation efficiency of 90%, as compared to the 64% efficiency of the traditional furrow irrigation system. As a result, average water requirements were reduced by 40% between 1975 and the end of the 1990s. At the same time, agricultural output increased twelve-fold.

The Israeli project to conserve urban water was conducted between 2002 and 2004, during which time water-saving devices were installed in the public buildings of local authorities, including schools, kindergartens, local youth culture centers, municipal/local council buildings, etc.

- Only devices labeled with the Blue Mark (a water conservative labeling system) were installed;
- The installation operation was applied across all municipalities and local councils in Israel;
- About 62,000 dual flushing cisterns, 3,000 waterless urinals, 105,000 sink- and shower-saving devices, and 2,000 garden volumetric valve and local irrigation controllers were installed.

Based on data collected from 30 authorities, it has come to be known that the average water saving in a building where water-saving devices were installed is 25%, (Lev 2012).

Between 1996 and 2000, and again in 2002 and 2004, the city of Santa Fe and the surrounding area experienced very dry years. With the intent to reduce per capita water use, the city of Santa Fe instituted emergency water conservation measures that included, among other provisions, restrictions on residential and commercial outdoor watering, as well as water-saving measures in commercial and public spaces. Additionally, the city implemented a comprehensive conservation program, documented in the Water Conservation and Drought Management Plan of 2005, which combined a number of different elements including specific water conservation requirements, water rate conservation incentives, water use audits, water offsets for new development, and general conservation education for the public. The primary, non-emergency water conservation measure, implemented in 2003, required all new demand on the water utility to be offset by replacing high flow toilets with flush toilets of 6 liters (1.6 gallons) or less.

Santa Fe experiences a dry steppe climate with chilly, dry winters and hot summers, and has a high potential evaporation compared to precipitation. General results

TABLE 1.1
Technologies to Consider for Conserving Water in Urban Water-Using Devices (Savage 2009)

Category	BAT (Best Available Technology)	BPT (Best Practical Technology)	Application	Cost	Saving
Toilets and Urinals	Composting toilets and waterless urinals	High-efficiency toilets and waterless or high-efficiency urinals	ALL	Cost varies, starting at about \$200–\$300	The EPA estimates that replacing an older toilet with a WaterSense labeled model will, on average, save more than \$90 per year in reduced water utility bills, and \$2,000 over the lifetime of the toilets; this equates to a saving of 4,000 gallons of water (15,000 liters) per year. Waterless urinals are totally water-free, and so save 100% of previous water use from day one.
Clothes Washers	ENERGY STAR clothes washer	ENERGY STAR clothes washer	ALL, esp. laundry facilities	Cost varies, starting at about \$800 w/o installation	For residential-style models, this can save an average of \$50 per yearly utility bill and 68 liters of water per load (equivalent to one shower). For commercial-style clothes washers, utility bills can reduce by more than \$1,000 per washer over ten years.
Dishwashers	ENERGY STAR dishwasher	ENERGY STAR dishwasher	ALL esp. food service	Cost varies, starting at about \$400 w/o installation	An ENERGY STAR qualified commercial dishwasher can save a business approx. 90 MBtus of energy, and an average of \$850 per year on their energy bills. In addition, businesses can expect to save more than \$200 per year and 200,000 liters per year due to reduced water use. ROI is 1–2 years.
Gray Water	Gray water system for irrigation and toilets (sink to toilets)	Gray water system for irrigation	ALL	Might be more costly in urban areas	Depending on the size, these systems will usually pay off within 1-2 years.

from a study of these activities indicate that water use, with some exceptions, has declined in the city of Santa Fe across residential, commercial and community categories. Average annual water use across lot size fell by 31% between 1998 and 2007/2008. The distribution of water use within residential areas generally shows a relatively "normal" distribution, but with a "tail" on the high-use side. This indicates that a few residents in each subdivision use a disproportionately high amount of water. Water use generally declined across commercial categories. Categories that demonstrated particularly large downswings included restaurants and hotels: the average water use per seat for full-service restaurants dropped by 50%, whereas average water use per room for full-service hotels fell by 58%. Exceptions to the general declining trend in commercial categories included grocery stores, gas stations and limited-service carwashes, where the average annual water use went up by 13%, 43%, and 26%, respectively.

Results for community categories also demonstrated a generally declining trend. The fall in annual water use per acre in some public parks is particularly noteworthy, though the total average fell only by 16%. The decline in average annual water use per 100 students was greater in primary and middle schools at 34% and 48%, respectively, than in high schools at 2%. Average annual water use at places of worship without daycare also fell a substantial 67% since 1998 (Borchert et al. 2009).

The conservation results of using low-flow fixtures including low-flow showerheads and faucet aerators were evaluated in the city of Kashan. Kashan, with a population of 275,000, is located on the edge of the central desert of Iran and has suffered extreme water shortages in recent years, much like the rest of the Middle East. To evaluate the efficacy of water conservation devices (WCDs), 2 groups of 40 households were randomly selected as experimental and control groups. The fixtures were installed in the houses of the experimental group, and water consumption was measured over one month. Results indicate that retrofitting these fixtures reduces residential water consumption by about 22%. Projections of Kashan's future water demand and supply indicate that, by using these fixtures, Kashan residents can postpone the need for new water supply projects by up to 6 years. The cost–benefit ratio of this conservation measure for Kashan is estimated to be 5.8:1 (Maleki-Nasab et al. 2007).

1.4 CONCLUSION

Water resources are subject to severe over-use and shortages around the world. The wise and optimal use of water resources is an essential action, especially for urban water. Water is often transported to urban and rural areas through pumping, with high energy consumption; after treatment, this scarce and valuable water is pumped into vast water distribution networks. However, these networks are mainly leak-prone, and consumers and appliances alike are wasteful.

REFERENCES

Borchert, C., Lyons, D., Trujillo, A., (2009), *Water Use in Santa Fe*, Water Division City, Santa Fe, NM, USA.

Lev, Y., (2012), Israeli Experience in Water Saving in the Municipal Sector, State of Israel Water Authority, Tel–Aviv.

Maleki-Nasab, A., Abrishamchi, A., Tajrishy, M., (2007), *Assessment of Residential Water Conservation due to Using Low-Flow Fixtures*, Sharif University of Technology, Tehran.

Mudgal, S., Benito, P., (2009), *Study on Water Efficiency Standards*, European Commission, DG ENV, Paris.

Savage, M., (2009), Every last drop, water, and sustainable business, *Virginia*, USA, http://www.sustainabilityconsulting.com/.

United Nations World Water Development, (2009), Water in a challenging world, Report 3, United Nations, 318 p.

2 Water Flow Controllers/ Restrictors

Abbas Yari and Saeid Eslamian

It is calculated that a decrease in the optimum flow rate from 6 to 5 liters/min (a reduction of 1 liter/min) can reduce the total amount of hot and cold water used by 4.7 liters/min, or a decrease of 13%. Additionally, the amount of hot water used is reduced to 3.2 liters/time for a kitchen washing-up experiment or a decrease of 12%. By replacing an existing inefficient faucet with a more efficient one, the annual saving would be 40 m^3.

The advantages of water conservation devices are savings in utility costs and energy (electricity or gas), environmental protection, and assistance in overcoming water scarcity.

Domestic water heating is estimated to be the second-largest energy end-use for Canadian households.

By installing a flow regulator for a shower, the water flow is reduced from 15 liters/min to 9.46 liters/min. Over 10 years, this results in water savings to the amount of 166,200 liters.

Aerators add air to the outlet water in such a way that the consumer does not feel the insufficient flow but is in fact satisfied.

Water controllers/restrictors are installed either in the faucet body or in miscellaneous plumbing, but aerator is installed at the end of the line, into the tap outlet.

2.1 INTRODUCTION

Based on a Japanese study, it is calculated that a decrease in the optimum flow rate from 6 liters/min to 5 liters/min (a reduction of 1 liter/min) can reduce the total amount of hot and cold water used by 4.7 liters/min (from 39.7 liters to 35.0 liters) i.e. a decrease of 13%. Additionally, the amount of hot water used is reduced to 3.2 liters/time for a kitchen washing-up experiment (from 30.2 liters to 27.0 liters) i.e. a decrease of 12%, (Yabe et al. 2014).

Each liter per minute increase in the consumption of water by water-consuming appliances such as a tap, for a four-person family, results in an additional water consumption of 2,100 liters (2.1 m^3) annually per each device. Replacing an existing inefficient faucet with an average flow rate of 15 liters/min, with a tap that controls water consumption (that is, replacing it with a 3-star valve, according to AS/NZS-6400 standard, that has a 9 liters/min flow rate), would save 22 m^3 annually. By replacing this faucet with a more efficient tap again (a 6-star valve with a flow rate of 4 liters/min), the annual saving will be increased to 40 m^3. In the above estimates, it is assumed that the water from each tap is used for handwashing, teeth-brushing, dishwashing, and the washing of fruits or vegetables for 10 minutes per day.

2.2 ADVANTAGES OF WATER CONSERVATION DEVICES

The advantages of water conservation devices are savings in utility costs and energy (electricity or gas), environmental protection, and assistance in overcoming water scarcity. Urban water conservation also results in less energy consumption. In Japan, approximately 30% of primary energy is consumed for domestic hot water supplies. In the previously mentioned study in Japan, volunteers were asked to use the faucets during an average season (autumn) and a cold season (winter). It was confirmed that the conventional faucets with a straight flow have the largest flow of water at 5.6 liters/min, and that the water-saving faucets have smaller values within the range of 3.9–5.2 liters/min. According to the relationship between the optimum flow rate and the total amount of hot and cold water used, a change of 1 liter/min in the optimum flow rate can change the amount of hot water used to 2.6 liters/min, i.e. a water-saving effect of 10.6%. From the results of a questionnaire survey, it was concluded that the optimum flow rate is determined mainly by "the force of the water" and "the degree of the splash." Moreover, an extreme reduction of the optimum flow rate is believed to lead to user dissatisfaction, and a flow rate of approx. 4.5–5.0 liters/min, which is close to the appropriate flow rate ranges of the sample faucets with spray patterns indicating relatively high user satisfaction, is considered to be the most appropriate (Yabe et al. 2014).

In a study using flow trace analysis of 10 homes in the city of Seattle with an average occupancy of 2.6 residents, DeOreo and Mayer estimated the per capita daily consumption of hot water to be 95 liters. Household daily consumption of hot water was estimated at 247 liters. Approximately 40% of overall water use was attributed to hot water use. Seattle is a seaport city on the west coast of the United States that has a temperate marine climate, classified as an oceanic Mediterranean climate, with cool, wet winters and mild, relatively dry summers.

Faucets, showers, baths, and clothes washers had the highest per capita hot water use (Table 2.1). End-users that involved direct consumer behavior or preferences, such as baths or showers, resulted in more water consumption than appliances with pre-set water consumption patterns, such as dishwashers (DeOreo et al. 2001).

Domestic water heating is estimated to be the second-largest energy end-use for Canadian households, exceeded only by space heating. As shown in Figure 2.1, it accounts for approximately 22% of total household energy consumption. In 2002, approximately 22% of total Canadian residential greenhouse gas emissions were attributed to domestic water heating (DWH), an estimated increase of 13% since 1990 (Aguilar et al. 2005).

Assuming a two-person household, in which each person takes 300 5-min showers per year, by installing a flow regulator, the water flow is reduced from 15 liters/min to 9.46 liters/min. Over a period of 10 years, this results in water savings amounting to 166,200 liters. Assuming an inlet water temperature of 10 °C and a shower

TABLE 2.1
Household Hot Water Use from Flow Trace Analysis (DeOreo et al. 2001)

Category	Per Capita Hot Water Use (liters/day)	Household Hot Water Use (liters/day)	Percent of Total Hot Water Use in Each Category (%)	Percent of Overall Use that is Hot Water (%)
Bath	15.9	41.3	16.7	78.2
Clothes Washer	14.8	38.2	15.5	27.8
Dishwasher	3.4	8.7	3.6	100.0
Faucet	32.6	84.8	34.3	72.7
Leak	4.5	11.7	4.8	26.8
Shower	23.8	62.1	25.1	73.1
Toilet	0	0	0.0	0.0
Other	0.04	0.1	0.0	35.1
Indoor Total	**95.0**	**247.2**	**100.0**	**39.6**

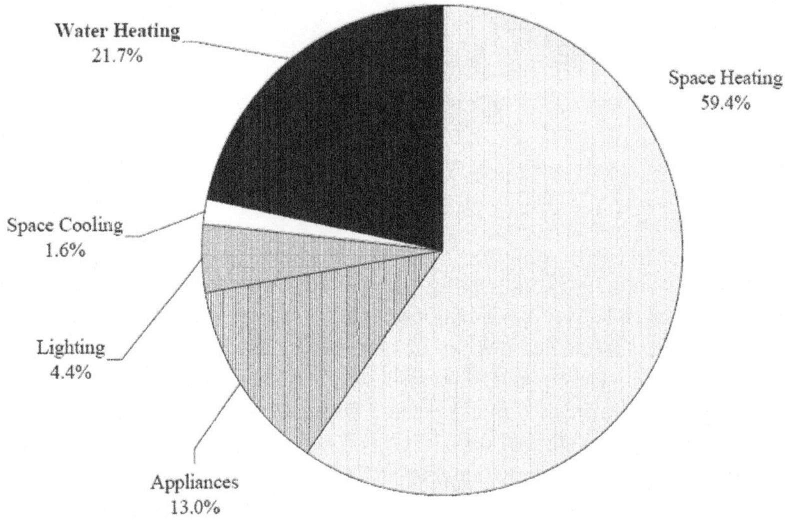

FIGURE 2.1 Canadian residential secondary energy consumption in 2002 by end-use (Aguilar et al. 2005).

temperature of 37 °C along with a specific heat capacity of 4.19 kJ/kg.K, this leads to an avoided thermal energy provision of 4831 kWh, (Berger et al. 2015).

2.3 AERATORS (END-OF-LINE FLOW CONTROLLERS)

Aerators (Figure 2.2) add air to the outlet water such that the consumer not only does not feel the insufficient flow of water but is also satisfied because aeration prevents

FIGURE 2.2 Aerators.

water from splashing and dispersing. The manufacturer of water conservation devices claim that these devices could reduce flow rate by up to 70% when installed at tap water outlets. These devices can be used in kitchen, lavatory, and laundry faucets, where they limit the capacity to nominal flow (The actual flow rate in the tap or shower stabilizes regardless of inlet pressure fluctuations and remains near-constant within the defined range of pressure).

The use of aerators has many benefits, including:

- Smooth and gentle water flow;
- The ability to control water spray by mixing water with air;
- A reduction in water and energy consumption relating to hot water production.

Without aerators, water tends to flow in a completely "pure" pattern from the appliance, but with the use of a modern aerator to add air, the water flow from the system is divided into small droplets, and the flow of water is automatically maintained, to the extent that it both needs and is expected to remain constant. By mixing air and water, the flow of water appears more intense, and the water-air mixture creates a

FIGURE 2.3 Aerator easy assembly equipment (US EPA WaterSense 2012).

sense of satisfaction in the consumer. These components are easily installed on taps and hoses, and even home customers can install or replace these devices within a few minutes (Figure 2.3), (US EPA WaterSense 2012).

The use of such intelligent and environment-friendly tools does not mean an interruption in water flow; instead, the required water is sufficiently available for washing dishes, bathing, washing hands and face, or brushing teeth.

In a study in Saudi Arabia, some ablution tests were conducted by three normal people and five traditional ablutions. Various types of water-saving devices, including self-closing taps, auto-faucets, aerators, and showerheads, were tested. About 41% of the water consumed during ablution was saved using aerators(Figure 2.4). However, about 35% can be conserved by using self-closing taps, and the auto faucet also saved 20% (Al-Rumikhani 2001).

2.4 FLOW RESTRICTORS (IN-LINE FLOW CONTROLLERS)

The pressure rise in building plumbing systems will increase both the flow of appliances and water loss. Flow controllers/restrictors are in-line flow controllers that are made in such a way that, by increasing pressure, the flow rate does not increase but remains constant.

According to manufacturers' claims, these devices can reduce the flow rate of sanitary facilities by 60%. These devices can be used in shower hoses and other hoses in residential and public sectors. The main type of these devices consists of self-regulating O-rings that are mounted in front of some orifices. The device compensates for excess pressure by flattening the O-ring when the pressure is increased and reducing the cross-sectional path of the orifices to reduce pressure and achieve the desired flow rate, as shown in Figure 2.5(a), b. Water controllers/restrictors are

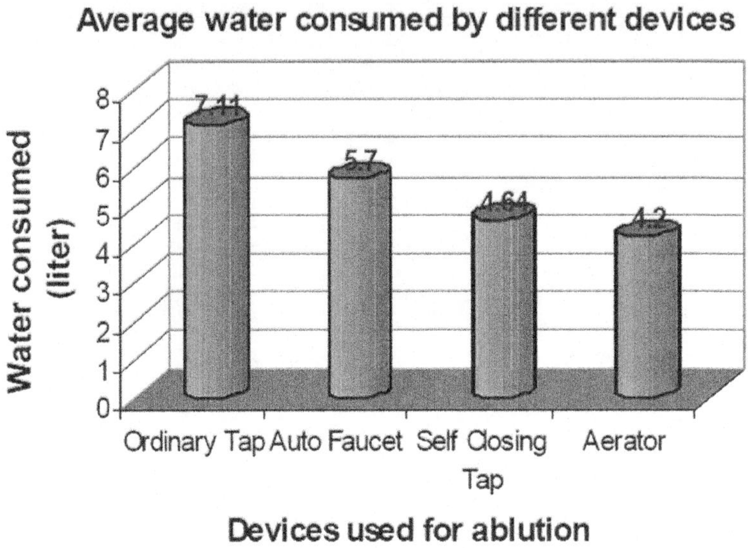

FIGURE 2.4 Average water consumed by different devices for ablution (Al-Rumikhani 2001).

FIGURE 2.5A Flow controllers/restrictor: washer type.

installed either in faucet body or miscellaneous plumbing, but aerators are installed at the end of the line into the tap outlet.

The Water Efficiency Labelling and Standard (WELS) scheme was launched by the Australian Government to help Australian consumers conserve water wisely. By using desalination and other technologies and policies, a comparison between the cost of water saved under the WELS scheme and other supply and demand options can be made, as per (Table 2.2). Based on this comparison, and given how much lower the cost of the WELS scheme is compared to other supply measures, it is likely that savings achieved under the scheme represent excellent value for money. The WELS scheme is cost-effective and achieves water-saving outcomes at a significantly lower cost than supply augmentation measures (Mollenkopf 2015).

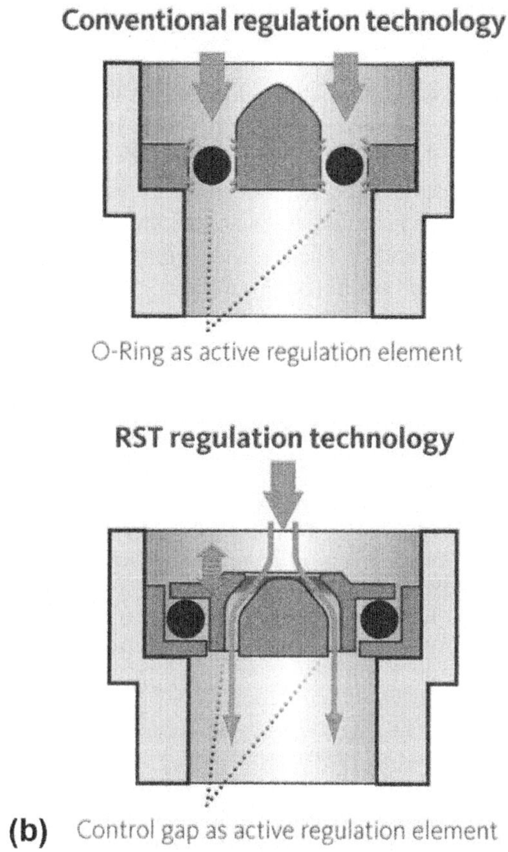

FIGURE 2.5B Flow controllers/restrictor: connector type.

TABLE 2.2
Comparison between Cost of Water Saved under the WELS Scheme and Other Supply and Demand Options (Mollenkopf 2015)

	Direct cost per kL of water	
Option	**Lower estimate**	**Upper estimate**
WELS Scheme	$0.08	$0.21
Demand management	$0.00	$1.45
Dams and surface water	$0.15	$3.00
Groundwater	$0.20	$1.58
Purchase of irrigation water	$0.63	$1.30
Rainwater tanks	$3.00	$5.60
Desalination	$1.15	$3.00

2.5 CONCLUSION

To reduce water use in showers and taps, the device can be replaced with a new one that meets local regulations and standards. However, a low-cost and straightforward solution to reduce the consumption by sanitary taps in the household and public buildings, as well as for dishwashing faucets, laundries taps, and showers, is to use flow controllers. These devices are small but efficient pieces that are well designed and carefully made. The devices are efficient, and the easiest and most inexpensive way to reduce water consumption. There is a wide range of products available to help save water in residential and commercial buildings, but flow controllers/restrictors, along with aerators, could be installed on existing fittings quickly and inexpensively to conserve water.

REFERENCES

Aguilar, C., White, D.J., Ryan, D.L. (2005), *Domestic Water Heating and Water Heater Energy Consumption in Canada, Canadian Building Energy End-Use, BEEDAC 2005–RP-02*, Natural Resources, Edmonton, Canada.

Al-Rumikhani, Y.A. (2001), *An Investigation of Water-Saving Devices: Performance and Saving Studies*, Natural Resource and Environment Research Institute, King Abdulaziz City for Science and Technology, Riyadh.

Berger, M., Söchtig, M., Weis, C., Finkbeiner, M. (2015), Amount of water needed to save 1 m^3 of water; life cycle assessment of a flow regulator. *Applied Water Science*. DOI: 10.1007/s13201-015-0328-5.

DeOreo, W., Mayer, P., Martien, L. (2001), *The End Users of Hot Water in Single-Family Homes from Flow Trace Analysis*, Aquacraft, Inc., Water Engineering and Management, Boulder, CO, USA.

Mollenkopf, T. (2015), *Second Independent Review of the WELS Scheme*, Commonwealth of Australia, Australian Government Department of the Environment, 2015, https://www.waterrating.gov.au/about/review-evaluation/2015-review.

Office of Water, U.S. EPA WaterSense. (2012), *WaterSense at Work: Best Management Practices for Commercial and Institutional Facilities*, Washington, DC.

Yabe, S., Otsuka, M. (2014), *Study on the Hot and Cold Water-Saving Effects of Various Types of Kitchen Faucets Using Different Spout Designs and Different Water-Ejection Modes*, Kanto Gakuin University, Kanazawa-ku, Japan.

3 Residential Water Use

Abbas Yari and Saeid Eslamian

There is tremendous untapped potential to increase water-use efficiency at homes, businesses, and the at the government level. At home, widespread adoption of water-saving appliances and fixtures for domestic users, along with the replacement of lawns with water-efficient landscapes, could reduce total residential water use by 40 to 60%. In the commercial, institutional, and industrial sectors, prior analysis has demonstrated that efficiency could be increased 30–60%. This would save an estimated 45,000 m^3 of water per year.

The performance of a toilet is a crucial requirement in evaluating its suitability. As such, premium models are exceptionally water-efficient and achieve excellent flush performance.

The old or non-standard showers used in most homes use up to 25 liters per minute; by using a convenient showerhead instead, even with a flow of 6 liters per minute, one can provide satisfaction for consumers.

A 5 kg washing machine with a 1-star rating consumes 150 liters of water per wash, but a 6-star washing machine with the same capacity consumes 25.2 liters of water per wash (only 16.8% of the total water use per 1-star device). It is observed that average water consumption is reduced by 47% when using a dishwasher.

Xeriscaping is the process of landscaping with slow-growing, drought-tolerant plants to conserve water and establish a waste-efficient landscape.

3.1 GENERAL

Water consumption sectors around the world vary widely from each other. Irrigation dominates water consumption for all regions worldwide, although its relative dominance varies based on climate and infrastructure, as shown in Figure 3.1 (Brauman et al. 2016).

Projected world water withdrawals from 2000 to 2050 by sector is illustrated in Figure 3.2; (OECD 2008). It shows that, although global agricultural water use is decreasing, overall water withdrawal is increasing due to an increase in industrial consumption. Agricultural water use is the still highest water consumer.

In the urban water supply, utilities are supplied to customers in a multi-division sector that usually includes single-family and multi-family households, commercial and industrial use, irrigation, and so on. There is a considerable variation in the composition of the customer base, from one municipality to the next. California's urban water use in 2005 was as shown in Figure 3.3. The highest consumption can be seen in residential outdoor use, followed by residential indoor applications (DeOreo et al. 2012).

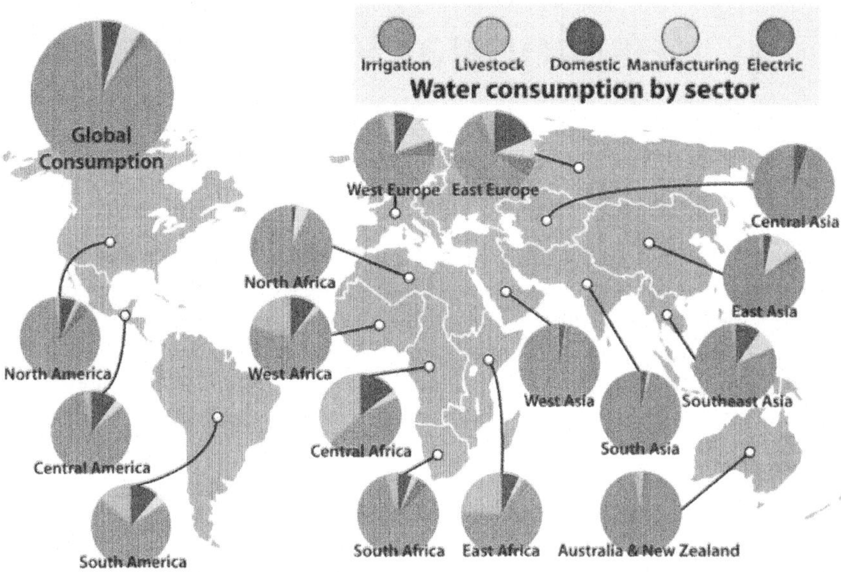

FIGURE 3.1 Water consumption by sector for regions worldwide (Brauman et al. 2016).

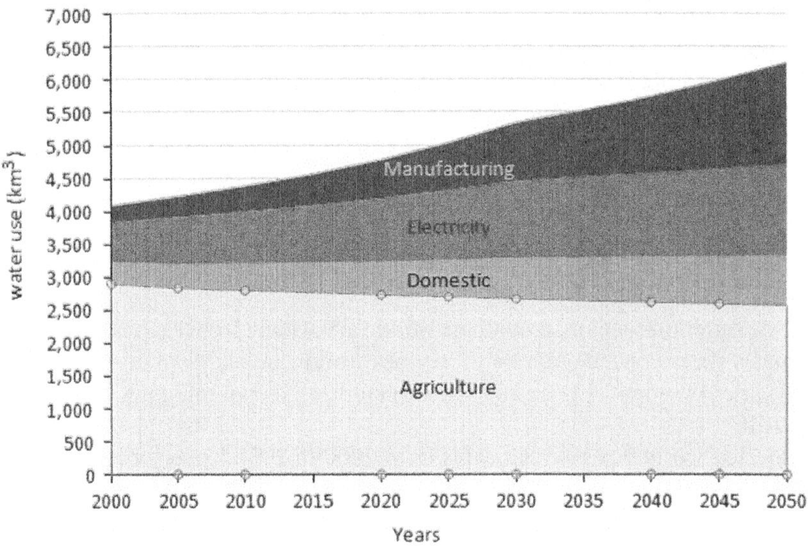

FIGURE 3.2 World water use by sector, 2000–2050 (OECD 2008).

There remains a tremendous untapped potential to increase water-use efficiency at home, in businesses, and in government. In the commercial, institutional, and industrial sectors, prior analysis has demonstrated that efficiency could be increased 30–60%. This would save an estimated 45,000 m^3 of water per year. At home, widespread adoption of water-saving appliances and fixtures, along with the replacement of lawns with water-efficient landscapes, could reduce total residential water use

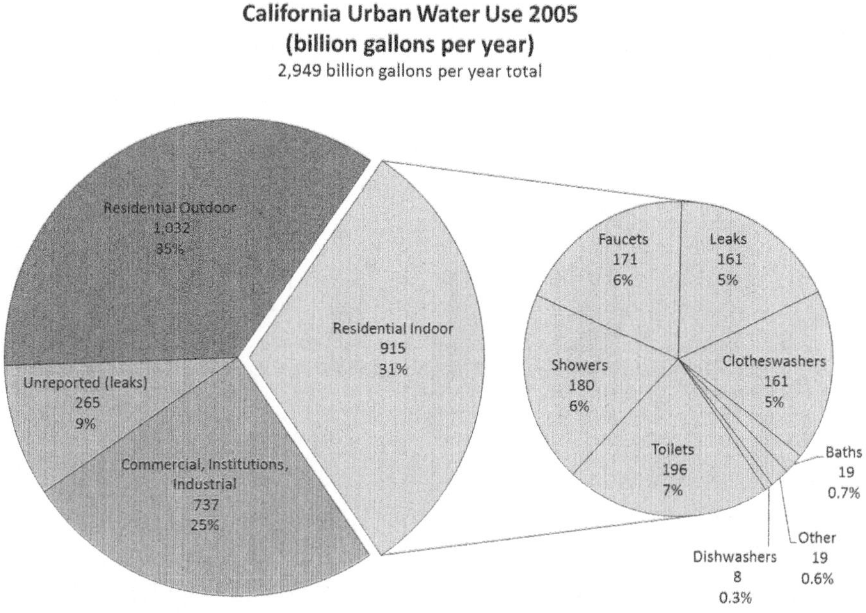

FIGURE 3.3 California urban water users (2005) (Christian-Smith et al. 2012).

by 40–60%. Improving water-use efficiency makes cities more resilient to drought, saves energy and reduces greenhouse gas emissions, lowers the cost of water treatment and new infrastructure, and frees up water to flow in our rivers and estuaries to benefit fish, wildlife, and recreational users, (Natural Resources Defense Council 2014). The shift in public perception towards water requires a renewed understanding of the relationships between the end user and the end users of water. Furthermore, despite successful demand management outcomes, an approach by many regulating authorities to reduce water consumption is often reactionary rather than proactive (Beal et al. 2011).

Significant indoor water savings are also available in the commercial, industrial, and institutional sectors. Limited data is available on water use and the potential efficiency savings for these sectors. The most recent quantitative assessment of commercial and industrial water conservation and efficiency potential in California was done by the Pacific Institute in 2003 by Gleick et al., and the authors' estimates have been adopted by state water planners. Using the estimates from this report, along with updated data on water use, it can be estimated that commercial indoor water efficiency could be improved by 30–50%, and that industrial efficiency could be improved by 25–50%, as per Figure 3.4 (Natural Resources Defense Council 2014).

Water use in toilets, showers, and basins often forms a large component of water use at commercial properties. The average daily demands of these conveniences can be as high as 155 liters per person in commercial and institutional settings. Savings of 25–30% are readily achievable in sites not operating efficiently. Furthermore, installing water-efficient appliances in high-water-use amenities, as well as maintaining fixtures, can be very cost-effective (City West Water 2012).

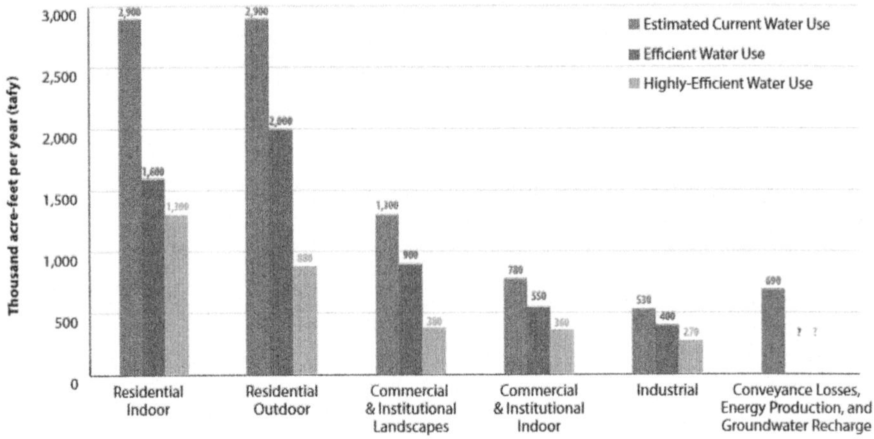

FIGURE 3.4 California's urban water conservation potential by sector (Natural Resources Defense Council 2014).

Between 1987 and 1991, Goleta city issued 15,000 rebates for high-efficiency toilets and installed 35,000 low-flow showerheads. Between 1983 and 1991, 2,000 new high-efficiency toilets were installed in new construction sites and remodels. Onsite surveys and public education efforts helped consumers to improve outdoor water efficiency; increased water rates provided extra incentive for consumers to reduce water use. Conservation and rationing programs, as well as rate increases, contributed to a 50% drop in per capita residential water between May 1989 and April 1990. Total district water use fell from 125 to 90 gallons (473 to 340 liters) per capita per day, a fall that was twice the original target of 15% (Whitman 2002).

As a result of being more efficient, technical changes in buildings allow for reductions in water demand. Table 3.1 presents potential savings through specific technological changes (Dworak et al. 2007).

Controlling and reducing the use of bathroom showers, taps, and kitchen faucets (which have a share of more than 60% of indoor water use) with small and inexpensive devices can be easily achieved. In homes and places where aflush tank is installed, the use of dual-flush siphons will also help to significantly reduce water consumption. The intelligent selection of water conservation devices (WCDs) can play an essential and influential role in the success of research, studies, and action. In terms of water efficiency and savings, a study on water-efficient standards in July 2009 by the European Commission's Directorate-General for the Environment estimated that showers and taps contribute 33% and 10% of average household water use, respectively. Replacing all standard domestic equipment (taps, toilets, showers, baths, washing machines, dishwashers, and outdoor consumers) with water-efficient products would result in an overall 32% decrease in domestic water consumption, or 40,716 liters per year for a European Union (EU) household (The European Water Label 2015).

The Loire-Bretagne water agency in France conducted a study with the objective of estimating current water consumption and potential water savings in various types

TABLE 3.1
Typical Water Saving Devices (Dworak et al. 2007)

Equipment	Description	Water-Saving
TAPS		
Taps with air devices	Introduction of air bubbles into the water increases its volume but results in less flow, maintaining the same effect	Flow reduction of around 50%
Taps with thermostats	Keeps the selected temperature	Reduction of around 50% of water and energy
Taps with infrared sensors	Water is dispensed when an object is underneath	Reduction of 70–80% of water
Electronic taps, or taps with buttons, that produce a timed length of flow	Water running for a limited time	–
TOILETS		
Toilets	A command for 6 liters per flush	–
Double-command toilets	A command for 3 liters per flush	–
Waterless or vacuum toilets	No water used	Reduction of water use by 50 liters per capita per day
WATER-SAVING DEVICES FOR OLD EQUIPMENT		
Device to mix water and air for taps	Increases the volume of water	Reduction of around 40%
Button to interrupt toilet flush	(reduction of flow)	Reduction of around 70%
Device to limit shower flow	(reduction of flow)	Reduction of 10–40%
Dishwasher	Decreases the volume of water used from 20 liters per use to 15 liters per use	Reduction of around 25%
WASHING MACHINES		
Washing Machines (~7kg load)	Decreases the volume of water used from 80 liters per use to 45 liters per use	Reduction of about 44%

of public infrastructures: educative buildings, sports facilities, hospitals, administrative buildings, public gardens, etc. The main results are summarized in Table 3.2. The sources of the data is feedback on municipalities' experiences, experts' estimations, scientific papers, and surveys (Dworak et al. 2007).

A major pilot project was initiated in seven cities in Brittany because of significant quantitative and qualitative problems with local water resources (e.g. water stress, droughts, and agricultural pollution). The cities involved were Brest, Lorient, Pontivy, Quinter, Rennes, Morlaix (St-Martin-des-Champs, France) and Vannes, corresponding to a total population of around 800,000. The project included a wide range of actions including an information campaign (users and professionals), letters to domestic users, tests on and installation of various types of water-saving

TABLE 3.2
The Saving Potential of Public Buildings in the Loire-Bretagne Basin (Dworak et al. 2007)

	Consumption of Reference	Saving Potential	Source
Primary school	3 m^3/child/year	20%	Lorient Pontivy, Brest, Douarnenez, Lannion, Perros, Guirrec
College	General: 3.6 m^3/student/year Professional: 6.1 m^3/student/year	18%	Conseil regional de Bretagne
Student housing	46.7 m^3/bed/year	30%	CROUS Aquitaine, Eco-Campus
Stadium (normal size)	1.000 m^3/year for equipment use 2.000 m^3/year for irrigation	20%	Surveys CNFPT Midi Pyrenees 2002, AIRES 1998, Report L.Cathala
Gymnasium (normal size)	800 m^3/an	15%	
Public swimming pools	0.33–0.42 m^3 /visitor	no data	
Hospital	100 m^3/bed/year	0%	Water agency data, experts
Administrative buildings	14.3 m^3/position/year	20%	Water agency data

equipment, and investigations of leakage in both the public distribution system and private households. Table 3.3 shows the significant quantities of water saved as a result of these actions (Lallana et al. 2001).

Denver Water has brought water savings to students in partnership with the Auraria Campus. The multi-year project included WaterSense-labeled showerhead retrofits. The utility also launched a water challenge in 2014 with condominium associations and apartment buildings that have high water bills. Eight buildings installed WaterSense labeled toilets, aerators, and showerheads. Large apartment buildings were easier to target than individual condos. The most significant challenge was getting enough people on board as many of the properties were individually owned units. Sometimes showerhead performance is questioned, and staff members suggest that the building install a labeled model in a common area like the pool or exercise room first to gather feedback. Denver Water also encourages building supervisors to make showerhead swaps part of their maintenance routine; for example, swapping out showerheads when they inspect the smoke detectors or check door locks. In the six months following the shower switch, Kennesaw State University (KSU) in the United States saved 2.5 million liters of water, or about 28% of the water used in the dorms, and about $6,500 in water bills. Since Cobb County paid about $9,500 for the showerheads, this retrofit at another university could potentially pay for itself in just one year (WaterSense 2015).

TABLE 3.3
Brittany Case Study: Water Savings Achieved (Lallana et al. 2001)

Equipment	Location	Investment Cost	Volume Saved	Percentage of Water Saved
Water-saving equipment for municipal irrigation	Brest	€2,000	43 m^3/week	About 62%
Water-saving equipment and leakage detection in individual schools	Brest	–	2.96 m^3/year/pupil	51%
	Lorient	–	5,500 m^3/year	79%
Meters and water-saving equipment in community halls	Pontivy	–	About 1,600 m^3/year	50%
Water-saving equipment in apartments	Rennes (43 apartments)	–	29 m^3/apartment/year	–
	Vanes (47 apartments)	€360	–	30% of toilet use
Water-saving equipment in swimming pools and leakage reduction	Rennes	€180	–	14%
	Morlaix (St-Martin-des-C hamps)	€560	2,340 m^3/year	30%
Detection of leakage in network	Moriaix (St-Martin-des-C hamps)	€3,800	1,300–1,800 m^3/year	–

Residents of San Diego County could get up to $2.75 per square foot ($29.6 per m^2) of grass that they replace with sustainable landscaping. It is advised to replace grass with plants that are recognized as moderate, low, or very low water-use type plants. The WUCOLs (Water Use Classifications of Landscape Species) reference guide is a useful resource for identifying plant watering requirements. Recently, the program has some additional requirements: rainwater harvesting and irrigation modifications are required, and synthetic turf is not allowed. Other requirements include mulch coverage around plants and a set number of plants per square foot (City of San Diego Public Utilities 2017).

The turf conversion rebate program by the Southern Nevada Water Authority (SNWA) will rebate customers $2 per square foot ($21.5 per m^2) of grass removed and replaced with desert landscaping, up to the first 500 m^2 converted per property, per year. Beyond the first 1,500 meters, the SNWA will provide a rebate of $1 per square foot ($10.76 per m^2). The maximum award for any property in a fiscal year is $300,000. The city of Hays, located in the county seat of Ellis County, Kansas, in the United States, offers a $1 per square foot ($10.76 per m^2) rebate for the conversion of maintained, irrigated, cool-season turfgrass to a more water-efficient, drought-tolerant turf grass or landscaping (Tillman and Bryantet 1988). Hays sits near the convergence of a humid continental and temperate semi-arid climate. It typically experiences hot summers with variable humidity, and cool winters, due to its geographic location at a climatic boundary.

3.2 DOMESTIC WATER USERS

Water is used in domestic buildings for a variety of purposes. The share of each consumer depends on a variety of factors, such as social, economic, and climate factors, and on the operating condition of the consumer products and how they are designed and manufactured. The per capita consumption of household customers in Abu Dhabi is presented in Table 3.4. As can be seen, the volume of water consumed for each consumer varies widely depending on ethnicity: Arab, Asian, Emirati or Western (Dornier Consulting 2014).

Household consumption structure varies largely across different studies. Matos et al. (2013) showed that in the north of Portugal, on a daily basis, it follows the following distribution: kitchen (38%), shower (26%), toilet (14%), taps (12%), clothes washing (8%), dishwasher (2%). Some other studies that present a household's consumption structures, in different conditions, are:

- Beal et al. (2011): shower (29.5%), clothes washer (21%), taps (19%), toilet (16.5%), leak (6%), irrigation (5%), dishwasher (2%) and bathtub (1%);
- Willis et al. (2013): shower (33%), clothes washing (19%), taps (17%), toilet (13%), irrigation (12%), bathtub (4%), dishwasher (1%), leak (1%);

TABLE 3.4
Per Capita Use by the Regional Group in Combined Monitoring Periods (Dornier Consulting 2014)

	Regional Group			
	Arab Expats	Asian Expats	UAE Expats	Western Expats
Parameter	Mean	Mean	Mean	Mean
Total Consumption lpcd	263.7	485.8	303.2	355.3
Landscape lpcd	79	269.6	58.8	199.2
Domestic lpcd	157.2	189	219.6	139.6
Bathtub lpcd	2.2	3.4	1.1	4.2
Clothes Washer lpcd	17.1	17.7	32.4	18.5
Dishwasher lpcd	0.5	0.4	0.2	1.2
Faucet lpcd	61.4	80	89.6	46.1
Leak lpcd	27.5	27.3	24.8	16.5
Other lpcd	0.05	0.03	0.02	1.75
Shower lpcd	38.1	49.1	57.4	37.1
Toilet lpcd	41.7	41.5	42.9	32.9

- Coghlan et al. (2003): bath and shower (33%), washing machine (27%), toilet (21%), taps (16%), other (3%);
- Almeida et al. (1999): bath and shower (28%), toilet flushing (31%), washing machine (16%), bathroom taps (13%);
- André et al. (1999): toilet flushing (31%), bath and shower (17%), washing machine (8%), dishwasher (0.3%) (Jorge 2014):
- MalekiNasab et al. (2007, Kashan, Iran): dish and clothes washing (24.7%), shower (26.1%), irrigation and outdoor washing (24%), toilet (11.5%), taps (10.5%).

Water use also related to the household family size and characteristics as shown in Figure 3.5, (Willis et al. 2013).

Figure 3.6 illustrates the average daily use of each of these end users in all income groups in Duhok, Iraq. Apparently, water consumption was directly related to family income (Hussien et al. 2016).

Until now, the most significant residential end use study conducted in North America was the Water Research Foundation's 1999 report, Residential End Users of Water (Mayer et al. REU1999). The WRF's new report, Residential End Users of Water, Version 2 (DeOreo et al. REU2016), provides an updated and expanded assessment of water use. It includes more varied study site locations, hot water use data, more detailed landscape analysis, and additional water rate analysis. The new study identifies variations in water use by each fixture or appliance, providing detailed information and data on changes since the 1999 report. Looking to the

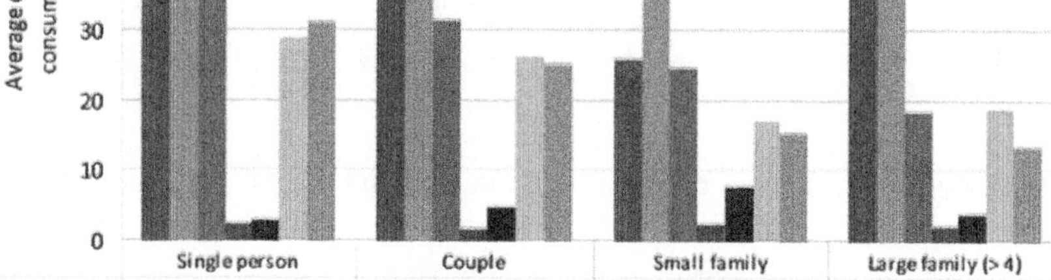

	Single person	Couple	Small family	Large family (> 4)
Clothes washer	45.1	36.3	26	35.9
Shower	61.8	56.1	45.6	41.3
Tap	38.0	31.4	24.7	18.6
Dishwasher	2.3	1.6	2.5	2.1
Bathtub	2.8	4.9	7.6	3.9
Toilet	28.8	26.2	17.0	18.7
Irrigation	31.2	25.2	15.3	13.5

FIGURE 3.5 The Relationship between Household Characteristics and Water End-use Consumption (Willis et al. 2013).

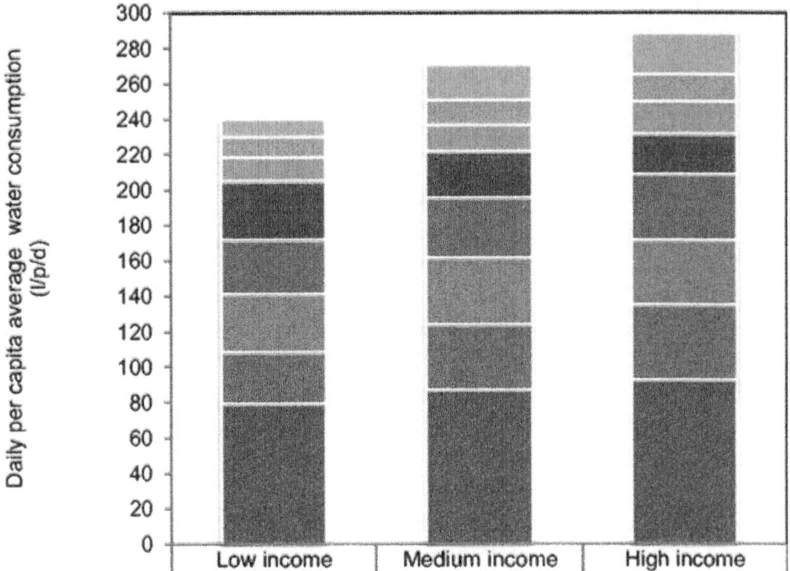

	Low income	Medium income	High income
Swimming pool	0	0	0.19
Bath	0	0	1.36
Vehicle washing	1.4	1.65	0.53
Garden watering	10.38	20.09	23.3
House washing	11.18	14.23	15.41
Cooking and drinking	13.2	14.85	18.33
Toilet flushing	32.99	25.45	22.51
Laundry	30.91	33.99	37.14
Dishwashing	32.98	37.98	36.69
Shower	28.74	36.67	42.31
Taps	79.43	87.27	92.59

FIGURE 3.6 Impact of Per Capita Monthly Income on Water End-Users in Duhok (Hussien et al. 2016).

future, the study evaluates conservation potential and includes predictive models to forecast residential demand. The decline in water use across the residential sector, even as populations increase, poses new challenges for water utilities. Information on single-family home water consumption is significant for utility rate and revenue projections, capital planning (water supply and infrastructure needs), daily operations to provide water, water efficiency programs, and more.

Based on the research, residential indoor water use in single-family homes has decreased. The average daily water use per household has declined 22%, from 177 gallons per household (gphd), or 670 liters per household (REU1999), to 138 gphd, or 522 lphd (REU2016). Per capita average water use has dropped 15%, from 69.3 gpcd, or 262 lpcd (REU1999), to 58.6 gpcd, or 222 lpcd (REU2016). In REU1999, a household averaged 2.77 people, and in REU2016, a household

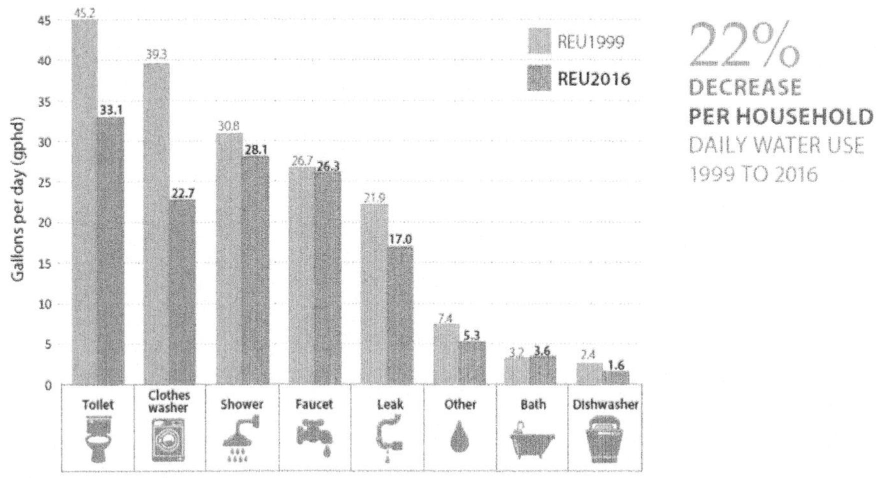

FIGURE 3.7 Average daily indoor water use per household REU1999 and REU2016 (DeOreo et al. 2016).

averaged 2.65 people. The improved water efficiency of clothes washers and toilets accounts for most of the decreases in indoor use, as shown in Figure 3.7 (DeOreo et al. 2016).

Average daily indoor water use per capita (REU1999 and REU2016) is as per Figure 3.8 (DeOreo et al. 2016).

Substantial indoor and outdoor conservation potential exists in the single-family sector. With 100%occurrence of higher efficiency devices, indoor household water

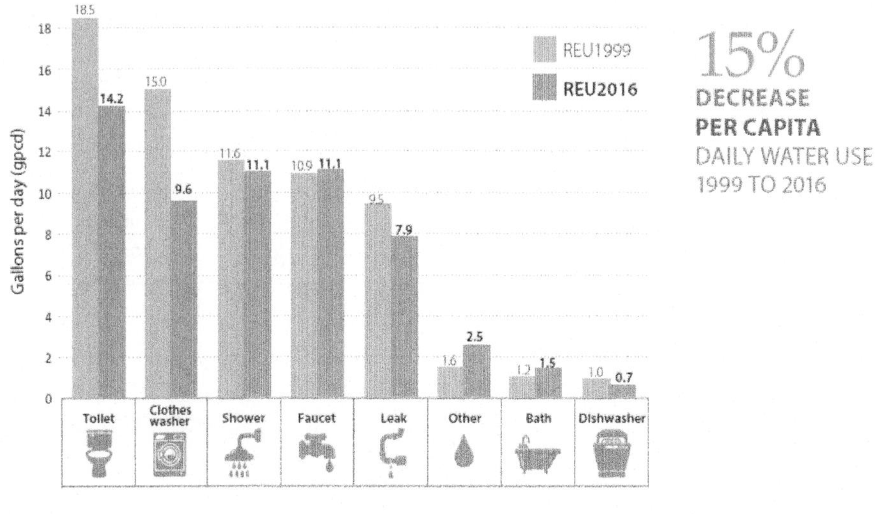

FIGURE 3.8 Average daily indoor water use per capita REU1999 and REU2016 (DeOreo et al. 2016).

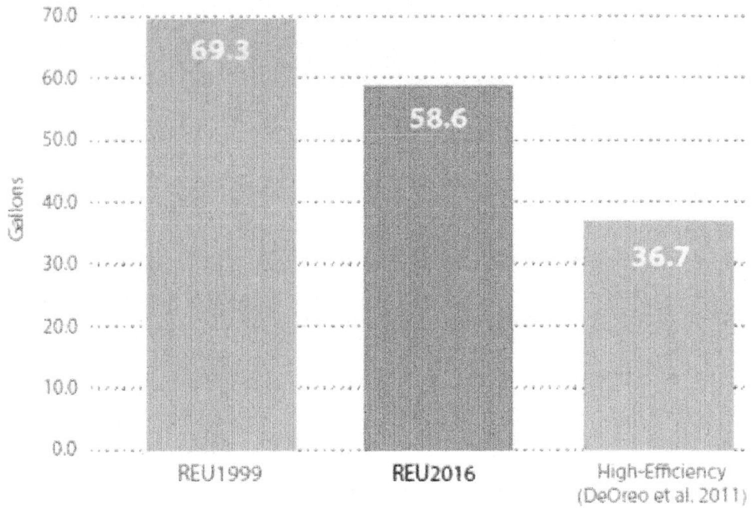

FIGURE 3.9 Indoor average gallons per capita per day (1 gallon = 3.785 liter) (DeOreo et al. 2016).

use could drop 35%or more, to below 150 liters per capita per day. Aggressive outdoor water conservation could reduce outdoor use even further, as per Figure 3.9 (DeOreo et al. 2016).

Consumption data for different Water-using Products (WuPs) present in residential buildings was collected for a study on water efficiency standards. Provided by the European Commission Directorate-General for the Environment, the data is summarized in Table 3.5 (Mudgal et al. 2009).

TABLE 3.5
Summary of Residential WuPs Consumption Data for EU Member States (Mudgal et al. 2009)

WuPs	Average water consumption per use	Frequency of use per person, per day	Average water consumption(l/household/day)	Share of household consumption (l/household/day)
Toilets	9.5	4.2	100	31%
Showers	50	0.85	107	33%
Taps	1.1	1	31	10%
Washing machines	60	0.6	37	11%
Dishwasher	20	0.57	11	3%
Baths	80	0.14	29	9%
Outdoor use	4.3	1	11	3%
Total	-	-	326	-

TABLE 3.6
Potential Household Water Savings from Water-Efficient Appliances (Mudgal et al. 2009)

	Standard product		water efficient product		Standard vs. efficient
WuP	**Litre/use**	**Litre/household/day**	**Litre/use**	**Litre/household/day**	**% reduction**
Toilets	9.9	100	5	53	53
Showers	50	107	40	86	20
Taps	1.1	31	0.8	24	23
Washing machine	60	37	41	25	32
Dishwasher	20	11	9	5	55
Baths	80	29	65	23	19
Outdoor	4.3	11	3.3	8	23
Total	-	**326**	-	**224**	**32**

If this overall water reduction potential by individual WuPs is disaggregated, it is evident that the highest savings can be achieved by promoting high-efficiency toilets (53%), as per Table 3.6. All products could potentially achieve reductions in water consumption around the 20% mark. It should be noted that, with the exclusion of toilets, the majority of the figures have been based on the water consumptions of the newest available standard products. Figures for consumption savings could be considered somewhat conservative in most cases, and actual savings may be much higher. By installing efficient water devices, the literature estimates a water savings potential of 29–41% per household (with the exclusion of outdoor WuPs) (Dworak et al. 2007). Using the data presented in Table 3.6, the results correspond to those in the literature, showing potential water saving in household consumption of approximately 32% (Mudgal et al. 2009).

3.3 COMMERCIAL, INSTITUTIONAL, AND INDUSTRIAL WATER USERS

Commercial, institutional, and industrial water uses are discussed in detail in Chapter 4 of this handbook. Concepts for conserving domestic water use in commercial, institutional, and industrial sectors are the same as described for residential users in this chapter.

3.4 RESIDENTIAL WATER-USE CATEGORIES

The water demanded by citizens in households and public places is consumed through a variety of activities. Case studies have shown that reducing water consumption to a reasonable level is possible with a low-cost program. In research, results indicated that sociodemographic, psychosocial, behavioral, and infrastructural variables all have a role to play in determining household water use. Putting aside factors that are out of the control of policymakers, such as household size and income, findings suggest that the significance of policymaking that promotes a culture of water conservation persists even when the environmental context changes. This promotion may be achieved through voluntary and mandatory actions that encourage ongoing water conservation behaviors and the installation of efficient appliances. Current low-level water restrictions, school-based education programs, and comprehensive campaigns that emphasize the precious and finite nature of water are strategies that can help to achieve this target. Securing water supplies in urban areas is a significant and perennial challenge for policymakers (Fielding et al. 2012).

Urban water consumption can be divided into two main categories: indoor and outdoor.

3.4.1 Indoor Water Use

All water consumption inside buildings, such as drinking, cooking, bathing, washing clothes and dishes, and brushing teeth, are indoor uses. The main appliances used for such applications are as follows:

3.4.1.1 Toilets

Water use in toilets can be reduced by:

- Installing water displacement devices, such as toilet dams, bags, or weighted bottles;
- Retrofitting flushometer (tank-less) toilets with water-saving diaphragms, which save 4 liters (20%) per flush;
- Replacing toilets with low-volume models.

Toilets can dump water to a total volume of 17 liters per flush, while low-volume toilets use only 6 liters per flush. An average saving of about 20% of the full water use in schools was possible through this conservation solution. Old toilet flushes consume up to 20 liters of (mostly refined) water per flush. Presently, dual-tank flushes with volumes of 6/4 liters and 4.5/3 liters are becoming popular. The design of dual-tank flushes may be attributed to Bruce Thompson of Caroma, Australia, who in 1980 developed a cistern with two buttons and two flush volumes (11 liters and 5.5 liters). The shape of the bowl had to be redesigned so that a low water flush could still efficiently remove waste. Thompson's design was piloted in a small town in southern Australia. The practice was successful with regards to water conservation, resulting in a piece of legislation that mandates dual-tank flushes in almost all new buildings. In a study at the University of Oregon in Portland, scientists' analysis revealed that building occupants are indeed utilizing the smaller flush, with reduced flush water use reduced by 32%. It is worth noting that despite this substantial water-saving, only 55% of building occupants expressed that they utilized the smaller flush on a regular basis, while21% indicated that the flip handle was difficult to use. This fact suggests that with the education of occupants and improved handle design, dual-tank flush water conservation in commercial buildings could be optimized (Harrison 2010).

Saving 1 liter per flush for a typical family would save about 3,200 liters per year. For an old toilet tank, the use of a plastic bag, a bottle, or a toilet tank dam reduces the amount of water used per flushing. In the city of Calgary in 2009, Emily Higginson at Veritec Consulting Inc. determined, firstly, if a product is practical and acceptable and, secondly, whether the use of a product will result in significant water savings. The purpose of this testing was to compare some of the different displacement devices to gain a general understanding of their effectiveness and usefulness. A variety of methods were tested, including a toilet dam, a Toilet Tummy (a bag filled with water), a Hippo Bag, a Dry Planet Bag, and a homemade Pop Bottle. The study showed that all displacement devices could affect the performance of a toilet, which means either these pieces of equipment dump less water or prevent a full-capacity refill. This may be disregarded if the device is installed in a toilet with powerful flushing, but if it is installed in a toilet with weak flushing, then the flush will be even worse. The testing verified that the small level water savings that could be achieved by introducing displacement devices are likely not worth the risk. Further, if the toilet experiences a leak for even a few days, this will invalidate any savings made. The study also found that none of these devices worked well (or even worked at all) in all toilet models; that is, while some worked better than others in specific models, the same displacement devices might in fact be problematic in other models. Although the displacement bag is a relatively inexpensive solution to promote water efficiency, it poses the risk of leaks and damage to homeowners. The researchers believe that the installation of efficient toilets is a far more reliable and efficient method of reducing water consumption within households (Higginson 2009).

Toilet flushing can account for up to 90% of water consumption in commercial buildings. Based on existing technology, several manufacturers offer WCs with a flush volume of 3 liters. State-of-the-art toilets, which use displaced air and water, produce a high-performance flush that requires only 1.5 liters of water.

Toilet performance is a crucial requirement in evaluating its suitability. To qualify the premium designation, toilet models have a target flush volume of 4.16 liters per flush (1.1 gallons) or less – most are rated at 4.0 liters per flush. Furthermore, all premium models must achieve a Maximum Performance (MaP) score of at least 600 grams and be certified to the WaterSense specification for tank-type toilets. As such, premium models are exceptionally water-efficient and achieve excellent flush performance. The Maximum Performance (MaP) testing project was developed in 2003 in order to identify how well popular toilet models perform using realistic test media. The testing protocol, cooperatively developed by water-efficiency and plumbing fixture specialists in the U.S. and Canada, incorporated the use of soybean paste as a test medium, closely replicating the "real world demand" upon fixtures.

However, achieving high MaP scores and being certified to WaterSense do not, in themselves, make these toilet models suitable for all types of installations, specifically in some non-residential regions where external factors dictate whether or not they will perform satisfactorily as a component of the total building's waste removal system. For example:

1) Without sufficient supplementary water (from clothes washers, showers, baths, dishwashers, process water, food service operations, sinks, etc.), 1.06 gallons (4 liters) may be insufficient to move solid waste entirely through

the building drain pipes. While extra water is typically available in residential installations, this is not necessarily the case in non-residential applications where toilets and urinals are usually isolated from additional sources, and lavatory faucets in the toilet room are limited to a flow rate of only 1.9 liters per minute or less;

2) A PERC (Plumbing Efficiency Research Coalition) study determined that flush volumes of 1 gallon (3.8 liters) and below in non-residential applications could lead to draining line blockage (depending upon other variables), and thus such volumes are not recommended;
3) The use of a single threshold (the maximum of both residential and non-residential users) assumes identical conditions in all buildings, which is not accurate. For example, non-residential buildings tend to have larger drainpipe diameters, less slope in pipes, longer pipe runs, and less supplementary flows – all of which negatively impact the flow of waste through building piping. Unlike residential toilets, non-residential fixtures are sometimes required to flush seat cover papers, paper towels, or excessive amounts of toilet paper. As such, the viability of maximum flush volume thresholds for non-residential applications should be separated from those of residential. While most single-family residential users may be suited to the 1.06 gallons toilet models, commercial and multifamily residential applications may not.
4) The use of a single threshold equal to the maximum of both new and existing building structures fails to recognize that the built environment was constructed mainly to code criteria developed well over a half-century ago. It is virtually impossible to alter existing building drainpipes in the built environment to accommodate flush volumes of 1.06 gallons (4 liters) and less. On the other hand, new construction could, in some cases, be explicitly designed to provide very low flush volumes and flow rates. For these reasons, MaP maintains that it is inappropriate to mandate a single maximum of 1.06 gallons (4 liters) flush volume equally for all residential and non-residential toilet fixture installations (Maximum Performance, MaP 2016). People in Muslim countries mostly use the water hose for ablution. In some field studies in Iran it is observed that using a reduced flow hose/valve that reduces the flow, does not meet the satisfaction of the users. Most of the reduced flow hoses have been destroyed by public users, especially in 6 lit/min hoses and less and also with a weak focused water jet.

3.4.1.2 Showers

The flow rate for a shower is expressed in liters per minute, the same as some other water consumers. Old or non-standard showers used in most homes use up to 25 liters per minute; by using a convenient showerhead, even a flow rate of 6 liters per minute can provide satisfaction for users. Most international regulations set the maximum water consumption to 8 liters per minute for showers.

In developed countries, the shower is the most significant indoor water use followed by faucets and toilets. Showerheads are one of the least expensive conservation measures available. Installing water-efficient showerheads and encouraging

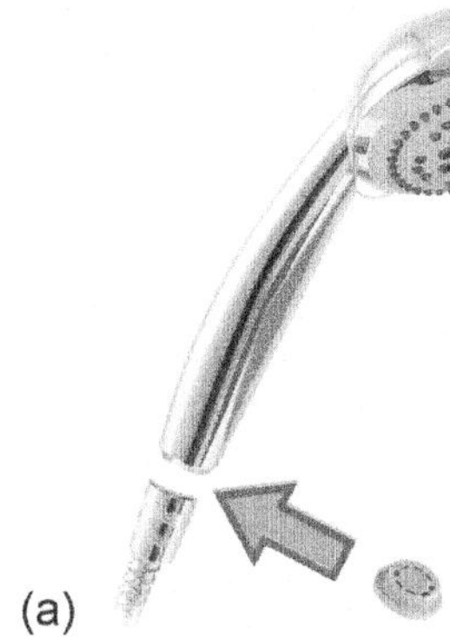

FIGURE 3.10A Washer flow control is attached to the showerhead.

shorter showers is one of the easiest ways to save water and energy and to reduce the overall cost of water and energy bills.

A shower's flow could be restricted through the use of flow controllers-restrictors, both washer and connector types, as shown in Figure 3.10. Washer flow control is attached inline on the showerhead in Figure 3.10a and connector flow controls are installed on the showerhead's hose or directly on the showerhead, as in Figure 3.10b.

Two types of water-saving showerheads are available. The most common heads save water through the aeration of the water, which is the process of infusing air into the water flow. This will cut water consumption by up to 50%, as some of the water is replaced by air bubbles. The disadvantage of this type of shower head is reduced water pressure. Showerheads with new pulsating technology provide a similar water saving but without any loss of water pressure. This is achieved by supplying water at high pressure, but instead of a continuous flow, the water is turned off and on 30 times per second. A water-saving is achieved each time the water flow is turned off. The shower experience feels normal because the pulsating is fast enough for the

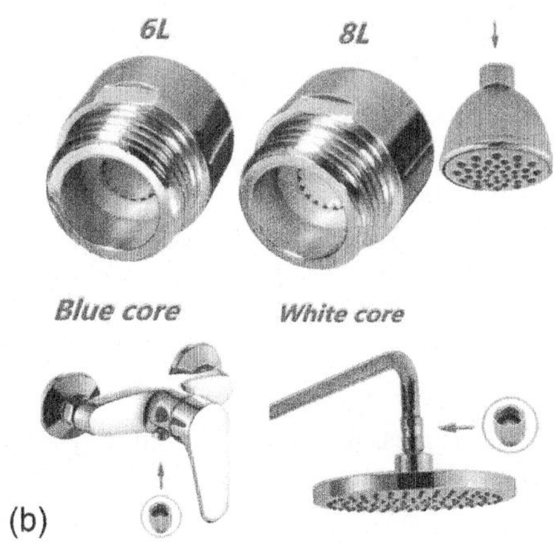

FIGURE 3.10B Connector flow controls are attached to the showerheads.

interruption in water flow to go unnoticed. Pulsating shower heads are more expensive than aerating showerheads. There are also some water-efficient showerheads equipped with a built-in pause handheld valve that allows the water stream to pause when soaping up for added water and energy efficiency during showering.

A summary of regular and energy-saving showerheads are outlined in Table 3.7. Low-flow showerheads also have fewer problems with regards to the calcification of nozzles and are less prone to dripping when installed appropriately, both of which can save money on maintenance and cleaning. When evaluating savings, it is essential to consider the number of people using the appliance. Savings will be much higher if people limit their shower time to a maximum of four minutes (City West Water 2012).

A study in Taiwan compared the showering profiles of behaviors classified as economic and squandering. The difference in showering period and water demands between economic behaviors (which averaged 4.2 minutes and 37.1 liters, respectively) and squandering behaviors (averaging 6.3 minutes and 56.9 liters) were about 2 minutes and 20 liters, respectively. The results of comparing ten different showerheads to identify the most comfortable one showed that the showerhead with a fine diameter of holes, a high velocity, a low flow rate, and an intermittent round flowing type is the most popular type. Based on the analysis, changing the water-saving showerhead from 6.5 liters per minute to 7 liters per minute, and improving the behavior in the economic process, could save more than 50% water demand (Lee 2014).

3.4.1.3 Faucets/Taps

Faucets are used by different users, such as basin taps, kitchen taps, hand/overhead showers, and bath and shower taps. During the study to evaluate high-efficiency indoor plumbing fixture retrofits in single-family homes in East Bay, California, it has been observed that faucets accounted for 12.1% of the total indoor water use during the baseline study period, out of which hot water had a 65.2% share. A useful way of evaluating faucet utilization is to calculate the duration faucets are utilized per capita daily. Observing satisfaction rating measures, 67% of the respondents would recommend the aerators to a friend, 10% said they would not recommend the devices, while 23% were not sure.

Faucets or taps are available in many configurations and designs. Taps can be manually opened and closed or be self-closing via mechanical or electronic actuators. They can be installed to deliver water to washbasins, bidets, or kitchen sinks.

Kitchen sink faucets are mainly used for washing fruits and vegetables and for dishwashing, as well as cooking. With the use of aeration devices or flow controllers-restrictors, the valve output can be limited to a certain extent.

Best practice for dishwashing by hand is to pour some warm water up to a third of the way into the bowl of the sink, then to add a few drops of dishwashing liquid and let the dishes soak in this solution. Thus, a faucet with a flow of 6 or 8 liters per minute is sufficient for washing dishes. A tap with the same specifications is appropriate for washing fruits and vegetables. Most national standards and regulations appointed 8 liters per minute as the maximum flow for a kitchen tap. Based on the Australian standard, AS/NZS 6400:2016, Class A is considered to be a sanitary valve with a flow rate of $4.5 \leq Q \leq 6$ liters per minute.

TABLE 3.7

Typical Showerhead Water and Energy Savings (City WestWater 2012)

Savings per Person (based on average shower time of 4.74 minutes and 0.77 showers per day)

Best practice flow rate (liters/min)	Typical flow rate (liters/min)	Kiloliters / year	Water US$/year	Electricity saving kWh	Electricity US$/year	Greenhouse gas-saving for electricity (kg $CO_{2\text{-}e}$)	Gas saving MJ (80% heating system)	Gas $/year	Greenhouse gas-saving for gas (kg CO2-e)
US$9.0	15.0	8.0	US$28.48	310.1	US$46.52	378.3	1,395.5	US$10.47	71.4

Washbasins constitute about 13% of household water used indoors. Some tests provided by an Iranian consulting engineers firm showed that a spray tap with a flow rate of 2 liters per minute could provide an acceptable level of satisfaction in public places such as mosques and schools. However, the 2 liters per minute spray could not satisfy the citizens in their workplaces and homes.

Automatic valves that allow water to be drawn off for use may be of the type that:

- Is manually opened, but which closes automatically giving a set period of flow. The period of the stream may be adjustable at the time of installation;
- Is electronically opened/closed. These values are actuated by a system that detects the presence of a user. Such systems might require a touching action or else might operate touchless (hands-free). The period of flow may be pre-set at the time of installation, or be constant while a user presence is detected (The European water label industry scheme 2015);
- Is proximity based. These sensor taps are electrically powered and do not require the user to touch them. However, proximity sensor taps are usually more appropriate for public buildings rather than households. They are more expensive and require a transformer, a power supply, or an extended life battery;
- Uses control devices to stop the water stream after a certain period or if they are no longer being used. Examples of control devices include:
- Sensors that pause the stream when a user leaves the sensor range (electronic opening and closing sanitary tapware);
- Time limiters that prevent the flow of water when the maximum flow time is reached (automatic shut-off valves);
- Low flow taps that reduce water consumption where the flow rate dictates the amount of water used. As taps are frequently used to provide water to fill containers, a reduced flow rate tap may be an inconvenience as the filling process may be prolonged;
- Automatic shut off valves are operated by pulling or pressing on a button that supplies a quantity of water depending on the time the tap is used and/or on the time delay time for its closing. The valve is operated by direct water pressure (Lev 2012).

Taps with water brakes, commonly known as 'click' taps, which are available in most manufacturers' product range. The workings of a typical click tap are shown in Figure 3.11.

For taps with water brakes, as the operating lever rises, the water flow increases. Water brakes usually are set to enable about 50% of the full flow to be achieved, although other percentages could be produced. Additional force to overcome the water brake is required to open the tap any further. Once overcome, the lever will move as easily as before towards full flow. While water brakes can be incorporated during the manufacture of the pillar and rotating taps; they are usually only fitted to mono-block mixer taps.

Click taps have the potential to save water in use, but they do have some limitations. Firstly, because they use ceramic cartridge technology (with smaller waterways

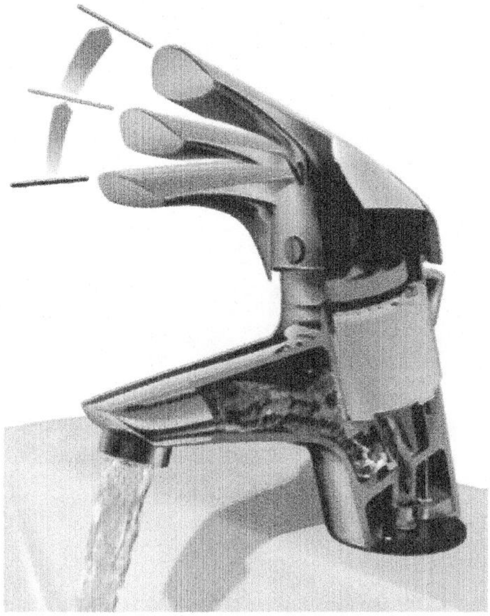

FIGURE 3.11 Section through a click tap with a water brake.

than traditional rising washer taps), they are only suitable for high-pressure systems. High-pressure systems are usually mains or pumped water fed and will deliver a dynamic pressure exceeding 1 bar (some "high pressure" fittings even require a pressure exceeding 3 bars to operate properly). Secondly, the extra resistance imposed by the design can be a problem for people with limited strength in their wrists. The use of a click tap with an additional flow restrictor may not have a cumulative impact as the click tap may no longer be able to decrease the peak flow by 50% (NHBC Foundation 2009).

3.4.1.4 Washing Machines

It is likely that the most important home appliance in facilitating modern human life is the washing machine. Considering the very high impact of the use of the modern washing machines, water authorities in many developed countries have prepared a list of high-efficiency washing machines and appoint discounts and rewards for those who use one of the machines on the list. These incentives are also provided for dishwashers in some cities, but due to the higher consumption of water and energy in washing machines, these incentives are less attractive for dishwashers.

Washing machines have experienced a marked increase in baseline efficiency, from 2006 where approximately 70%of the market was represented by products with a 2.5 WELS star or below, to now, where such machines represent less than 1% of the market (Figure 3.12). Products with a 4-star WELS or better have dominated the market since 2015 (Tom Mollenkopf 2015).

Many countries have national standards for the determination of water consumption and water labeling instructions for household electric washing machines. According to these standards, manufacturers and importers are obliged to label

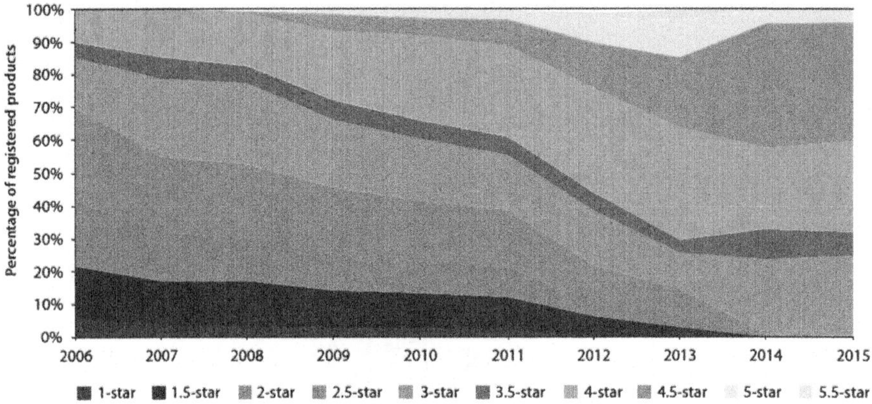

FIGURE 3.12 Percentage of WELS-rated washing machines 2006–2015 (Mollenkopf 2015).

washing machines according to these standards. For example, as per Australian standard, AS/NZS 6400:2016, a 5 kg washing machine with a 1 star rating consumes 150 liters of water per wash, but a 6-star washing machine with the same capacity consumes 25.2 liters of water per wash (only 16.8% of the consumption of the 1-star device).

3.4.1.5 Dishwashers

The most suitable option for washing dishes is using modern dishwashers. These machines are designed to cleanse all the dishes with minimal water and energy consumption. As water authorities in some developed countries pay a portion of dishwasher prices or offer a discount on water utility bills (water and electricity), consumers are encouraged to buy and install this equipment. New dishwashing machines use about half of the water that old dishwasher machines required. The energy consumption of these machines has also been remarkably improved.

Based on the Australian standard AS/NZS 6400:2016, a 1-star dishwasher consumes 18.5 liters of water per wash, but a 6-star dishwasher, use 7 liters of water per wash (only 38% of the total use of the 1-star dishwasher). According to an Australian estimate, replacing a 2-star dishwasher with a 4-star washer would cut utility bills by AU$770. 65% of these savings is due to a reduction in the cost of consumed energy (electricity or gas), and 35% is due to the reduced water consumption.

Table 3.8 shows a comparison between the water consumption of washing dishes with and without a dishwasher in four European countries. It is observed that average water consumption is reduced by 47% when using a dishwasher (Richter 2010).

3.4.2 Outdoor Water Use

About half of California's urban water use, equivalent to 5.2 billion m^3 per year, is outdoors. About 70% of outdoor use is residential, representing both single and multifamily homes. Commercial businesses and institutions account for the remaining

TABLE 3.8
Average Water Consumption in Households With and Without a Dishwasher (Richter 2010)

Country	With dishwasher (w/ DW) or Without a dishwasher (w/o DW)	Water Use per Item in a Liter (Arithmetic Average)
Germany	w/ DW	0.6
	w/o DW	1.0
Italy	w/ DW	1.0
	w/o DW	1.7
Sweden	w/ DW	0.8
	w/o DW	1.7
UK	w/ DW	0.6
	w/o DW	1.6
Total	w/ DW	0.7
	w/o DW	1.5

30% of outdoor water use. The highest rates of outdoor water use are in the hot, dry areas and in communities where water is inexpensive. In these areas, outdoor water use can account for up to 80% of the total use (Hanak et al. 2006).

Outdoor water use consists mainly of landscaping, as well as car or sidewalk washing, filling pools, operating fountains, washing yards, and sweeping away trash. Water reductions of 20–50% are estimated when mild to aggressive landscape conservation programs are used. If excess irrigation could be eliminated, the average outdoor water use would drop by 31 m^3 per house, or 16% (DeOreo et al. 2016).

Landscaping is one of the water users that should most likely be sacrificed in drought situations, which are sometimes called a watering ban. In 2016, after four years of drought in California, state officials moved to ban water-wasting habits. The adopted new mandatory water conservation rules, which affected millions of people, employed restrictions on a range of water uses, from how homeowners water their lawns to how restaurants and hotels serve their guests. It considered all outdoor water uses, such as spraying sidewalks with hoses, running sprinklers within 48 hours after measurable rainfall, and washing cars using hoses that do not have turn-off nozzles, by evaluating watering times and advising users to irrigate slowly and to minimize or eliminate evaporation, runoff, and overspray. The following cash rebates were offered to qualified applicants:

- Design: 50% of the cost of landscape or irrigation design services, up to a maximum of US$250 per site;
- Irrigation equipment: 50% of the cost of drip irrigation parts, sprinkler system efficiency retrofits, sprinkler nozzles, pressure compensating heads, pressure regulators;
- Water Wise plants, mulch, and permeable landscape material: 50% of the cost of water wise plants and mulch. Permeable landscape to include

synthetic turf, gravel, cobbles up to 4 inches, flagstone or similar with a minimum of 4 inches spacing. Planted areas must be covered with a 3-inch layer of mulch. Hardscapes and pathways, including decomposed granite, are not eligible;
- Smart irrigation controller: 50% of the cost of a smart irrigation controller. Smart irrigation controllers work on a simple principle: automatically adjust for weather changes and irrigate based on the needs of your plants;
- Pool covers: 50% of the cost of a pool cover, up to a maximum of US$300 per pool cover;
- Synthetic turf and permeable surfaces: 50% of the cost of artificial turf and other permeable surfaces;
- Laundry-to-landscape greywater: 50% of the cost of laundry-to-landscape greywater system parts.

Any combination of pre-approved irrigation equipment and planting costs may qualify for a one-time rebate of up to US$750 for single-family meter customers, and up to US$2,000 per meter serving the irrigated area for multi-family, commercial, and dedicated landscape irrigation meter customers (Goleta Water District 2018a). See Table 3.9.

There are many options and recommendations to conserve outdoor consumptions, like using harvested water, using drought-resistant grasses, not watering in the wind, cleaning drive and walk sides with a sweep, cutting watering back to once a week or once every two weeks by deep-watering the lawn, using a bucket to wash the car, and conducting irrigation after sunset, thoroughly, and less frequently.

The latest innovation in irrigation is the smart controller. By using smart irrigation controllers, irrigation devices give plants the right amount of water for the time of year, the climate, and the weather. Smart controllers are able to avoid overwatering and excessive run-off by scheduling the amount of irrigation based on the type of landscape and current weather. Water use efficiency programs are arranged on weather-based irrigation scheduling devices, a soil moisture-based smart system, and weather- and soil-moisture-based landscape irrigation scheduling devices.

Weather-based wireless irrigation controllers access weather data in real-time via the water user's Wi-Fi internet connection to automatically adjusts irrigation schedules. The controller can be operated, adjusted, and monitored via Wi-Fi using a web browser, mobile device, or at the controller. Some smart irrigation systems use "cloud" services that interpret hyperlocal, real-time weather data to control and monitor an unlimited number of controllers. Some other residential and commercial irrigation controllers operate based on weather conditions using onsite sensors.

Most of the control systems could be operated remotely by mobile apps and/or software. Apps for automated scheduling are optimized for water efficiency and landscape health by taking into account numerous variables including weather, seasonality, landscape characteristics, water budgeting, and vegetation.

Water authorities may consider some incentive programs that offer rebates on eligible models of Water Smart Landscapes. Others like the Center for Landscape Water Conservation, a resource for homeowners and industry professionals in New

TABLE 3.9
The Stage III Water Shortage Emergency Information Sheet (Goleta WaterDistrict 2018)

Stage III water use restrictions include:

- Outdoor landscape irrigation *remains* limited to no more than two times per week during early morning or late evening hours, and is *now* prohibited within 48 hours of measurable rainfall:
  - Manual watering (including with a sprinkler attached to a hose) is *now* only allowed before 8 a.m. or after 8 p.m., any two days per week.
- Use of fixed (i.e. installed) sprinkler systems must comply with the following updated schedule:
  - Residential properties may water Wednesdays and Saturdays, before 6 a.m. or after 8 p.m.
  - Commercial and institutional properties may water Tuesdays and Fridays, before 6 a.m. or after 8 p.m.
  - Public parks, athletic fields, and golf courses may *now* water no more than two days per week, before 6 a.m. or after 8 p.m.
- Hotels, motels, and other lodging are *now* required to post water shortage notices, and refrain from daily linen washing unless specifically requested by the patron.
  - Agricultural customers using overhead spray irrigation outdoors are *now* restricted to before 10 a.m. or after 4 p.m.

The following water use restrictions remain in effect:

- Hoses must be equipped with a shut-off nozzle.
- Direct application of water to sidewalks, pavements, open ground, or other hard surfaced area is generally prohibited.
- Washing buildings, dwellings or other structures is generally prohibited.
- Vehicles and boats may only be washed at commercial car washing facilities or with a hose equipped with a shut-off nozzle.
- Use of water in outdoor fountains, reflection ponds, and decorative water features is prohibited unless located on a residential property or home to aquatic life as of September 9,2014.
- Restaurants and other food establishments may not serve water unless specifically requested by the patron.
- Gyms, athletic clubs, public pools, and other similar establishments are encouraged to post water shortage notices at their facilities and promote shortened showers.

Mexico and West Texas, may feature a primary website and public networks and mobile app.

3.4.3 Xeriscape

In response to drought conditions, Denver Water created the term "xeriscape" in 1981 to promote water-wise landscapes. Xeriscaping is landscaping using slow-growing, drought-tolerant plants to conserve water and establish a waste-efficient landscape. A xeriscape is a garden or landscape that needs little supplemental water and xeriscaping is the process of landscape an area using water conservation principles. See Figure 3.13 for an example.

Xeriscapes do not have to be cactus or rock gardens. Often, they include other native and drought-tolerant plants. Xeriscape water conservation principles can be applied to different styles of landscaping. In arid and semi-arid climates, where water conservation

FIGURE 3.13 An example of Xeriscaping plant layering.

is important, xeriscaping is a wise choice. Xeriscapes can reduce the water needed for landscapes by 50–75%. This results in cost savings as well. Xeriscaping can also be used to reduce the amount of maintenance required for a beautiful yard or garden.

Xeriscaping incorporates seven basic principles of landscaping to achieve water conservation.

1. Planning and design: provides direction and guidance, mapping the water and energy conservation strategies, both of which will be dependent upon the regional climate and microclimate;
2. Selecting and zoning plants appropriately: plant selections and locations should be based on what will flourish in the regional climate and microclimate. Always group plants with similar water need together;
3. Limiting turf areas: this reduces the use of bluegrass turf, which usually requires a lot of supplemental watering. Consider substituting a turfgrass that uses less water than bluegrass;
4. Improving the soil: this enables soil to better absorb water and to encourage deeper roots;
5. Irrigating efficiently: use the irrigation method that most efficiently waters plants in each area;
6. Using mulches: these keep plant roots cool, minimize evaporation, prevent soil from crusting, and reduce weed growth;
7. Maintaining the landscape: keep plants healthy through weeding, pruning, fertilizing, and controlling pests (Office of Energy Efficiency and Renewable Energy 2018).

There are several sites and literature resources to explain firm water conservation landscaping. The Center for Landscape Water Conservation has published a mobile application to improve water-efficient landscape, called "Southwest Plant Selector". The app presents hundreds of xeriscape-friendly plants that can be used in gardening, landscaping, and horticulture projects to reduce water use.

Southwest Plant Selector contains expert-recommended xeriscaping landscape plants and is designed especially for New Mexico, El Paso, and surrounding areas. All plants in the app need little or no supplemental water and are typically both available and used in regional xeriscapes. Water users can select plants based on different parameters such as whether it grows in shade or sun, is deciduous or evergreen, its form, and its dominant bloom colors, along with a specific recommendation for each region of the county. The original data in this app was collected and organized by the New Mexico Office of the State Engineer and it is updated and edited by the Center for Landscape Water Conservation and the New Mexico Office of State Engineer in association with New Mexico State University (Sutherin et al. 2013).

It is important to use appropriate applicators for the plant materials and to fit the irrigated area for each zone. Many kinds of irrigation application devices can be used in landscapes. Irrigation zones may be differentiated in the design process because different application devices or different management is needed from one area to the next. Separating different kinds of application devices into different irrigation zones allows each zone to be operated in a way that is consistent with the application rate for similar application devices.

The two main categories of irrigation application devices are sprinkling type applicators or micro-irrigation applicators. Sprinkling application devices spread water in a broadcast manner over the whole area to be irrigated, mimicking rainfall, while micro-irrigation devices do not broadcast water over an entire area. To spread water evenly over an area, sprinkle application devices must have an overlap of their water streams. Micro-irrigation application devices, also called drip irrigation devices, apply water directly to the root zone areas of plants, minimizing the wetted area of the soil surface. They do not have overlapping wetting patterns (Berle et al. 2007).

Sprinkle applicators can be sprays or sprinklers. In spray applicators there is no movement of the water streams that form the pattern, or wetted area, of the spray, and there are no moving parts in the spray head. Figure 3.14 is a picture of spray applicators operating in a landscape. Sprinklers have moving water streams that broadcast water over an area that forms the pattern of the sprinkler. Sprinkler types include a rotor, impact, and rotator style movements of the water streams.

Figures 3.15(a) and (b) show a rotor and a rotator applicator, respectively. Most sprinkler or spray applicators can be housed in pop-up canisters that recede below the soil level when not applying water. They may also be placed on top of a riser, or a vertical length of pipe, to provide broadcast application above taller plants (Berle et al. 2007).

Micro-irrigation devices include drip emitters, inline emitters in the rigid lateral pipe, drip tape, micro-sprinklers, and micro-sprays. The most common micro-irrigation device in landscapes is the drip emitter, which is installed along a lateral pipeline laid on the surface of the ground. Drip emitters are positioned along lateral pipelines where needed to drip water over the root zone of individual plants.

FIGURE 3.14 Spray applicators with pop-up style heads.

Figure 3.16 shows a drip emitter that has been installed into a lateral pipe. Semi-rigid polyethylene pipe laterals have inline emitters embedded at regular intervals such as 12 inches within or along one side of the pipe. Drip tape is a flatter, more flexible polyethylene hollow "tape" with emitting devices evenly spaced along the tape. Drip tape and inline emitter piping can be buried or placed at the soil surface. Micro sprays and micro-sprinklers have low volume application rates similar to drip emitters, but the low volume of water is spread out over a larger area than with drip emitters. Micro-sprinklers are similar to micro-sprays except they have moving parts and water streams (Berle et al. 2007).

Figure 3.17 is a picture of a micro-spray that is not applying water. Neither micro-sprays nor micro-sprinklers are designed to have overlapping water patterns like other sprinkling application devices. These micro-sprays or micro-sprinklers are needed where soils have a low water-holding capacity and high permeability so that the water applied does not move below the root zone without adequately refilling a large area of the root zone. Some plant materials such as herbaceous perennials and woody plants are more amenable to drip irrigation, while turfgrass areas will be better suited to the broadcast application of sprinkler irrigation. The appropriate system for herbaceous annuals depends on their spacing and growth habit. Where there are steep slopes and plant materials other than turf grass, drip irrigation is a

(a)

FIGURE 3.15A Two rotor applicators with an over-lapping radius of throw (Berle et al. 2007).

(b)

FIGURE 3.15B Pop-up rotator applicator (Berle et al. 2007).

better system since it will apply water slowly and prevent runoff (Alliance for Water Efficiency 2018).

For turfgrass on steep slopes, low application rate sprinklers are preferred. The designer should choose applicators such that impervious surfaces are not irrigated. Spray applicators come in many shapes to accommodate difficult spaces. Both sprays and sprinklers can have arcs less than 360° to prevent the watering of impervious areas. Most sprinklers and sprays come in standard one-quarter, one-third, half, three-quarter, and full-circle arcs. There are also sprinklers and sprays with adjustable arcs for spaces that need more or less than the standard arc sizes.

Any irrigation device requires a certain operating pressure and flow rate. The operating pressure results from the static pressure at the upstream, or start, of the irrigation system minus the pressure losses in the delivery lines and equipment of the irrigation system. Static pressure is the pressure measured when there is no flow,

FIGURE 3.16 A drip emitter connected to a pipe lateral (Berle et al. 2007).

FIGURE 3.17 Micro-spray irrigation.

i.e. the pressure when reading a pressure gauge just behind or at a closed valve is the static water pressure at that valve.

To ensure adequate pressure for all applicators, analyze pressure losses due to distribution and topography for each irrigation zone throughout the system to ensure that applicators at locations with the least pressure are adequately pressurized. Careful choices in the layout of distribution lines can reduce the amount of pressure difference among the applicators within an irrigation zone, which will provide a more uniform distribution of water to the applicators.

Steep areas can be isolated for better management in irrigation zones separate from zones for flatter areas. This allows for better pressure control and better irrigation management within topography differences. Flat areas at the base of a steep area may receive some runoff from the steep area and will need less irrigation water. If the actual operating pressure is higher than appropriate for the applicators, the system uniformity will be compromised. Pressure regulators are needed where excess pressures would occur. High pressures are common at the base of steep slopes (Berle et al. 2007).

3.5 CONCLUSION

The traditional approaches to water supply and management, in combination with population growth and a raising of the economy, have led to the remarkable use of freshwater resources across most of the world.

The intelligent selection of water conservation devices (WCDs) can play an essential and influential role in the success of a water conservation action. Most WCDs are low-cost and straightforward solutions to reduce the consumptions of sanitary taps in the household and in public buildings. A series of water-saving technologies have been developed for household application, such as flow controllers, low consumption taps and showers, low consumption toilet flushes, low consumption washing machines, etc., that might allow for a reduction in individual consumption.

There remains a tremendous untapped potential to increase water-use efficiency at home, in businesses, and in government. At home, widespread adoption of water-saving appliances and fixtures, along with the replacement of lawns with water-efficient landscapes, could reduce total residential water use by 40–60%. Improving water-use efficiency makes cities more resilient to drought, saves energy, reduces greenhouse gas emissions, and lowers the cost of water treatment and new infrastructure.

REFERENCES

Alliance for Water Efficiency, (2018), *Drip and Micro-Spray Irrigation Introduction*, AWE, Chicago.

Almeida, M., Butler, D., and Friedler, E. (1999). At-source domestic wastewater quality, *Urban water*, 1, 49–55.

André, P. and Pelin, E. (1999). Brazilian National Program to prevent the waste of water, Economic analysis of domestic consumption, Office for Urban Development, Brazil, (in Portuguese).

Beal, C., Stewart, R., (2011), *South East Queensland Residential End Use Study*, Urban Water Security Research Alliance, Queensland, Australia.

Berle, D., Harrison, K., Seymour, R., (2007), *Best Management Practices for Landscape Water Conservation*, The University of Georgia College of Agricultural and Environmental Sciences and the U.S. Department of Agriculture Cooperating, Athens, GA, USA.

Brauman, K., Richter, B., Postel, S., Malsy, M., Flörke, M., (2016), *Water Depletion: An Improved Metric for Incorporating Seasonal and Dry-Year Water Scarcity Into Water Risk Assessments*, *Elem Sci Anth*, 4.

Christian-Smith, Heberger, Luch, (2012), *Lavatory Faucets and Faucet Accessories*, California Energy Commission, Sacramento, CA, USA.

City of San Diego Public Utilities, (2017), *Guidelines for City of San Diego Grass Replacement Rebate,* San Diego, CA, USA.

City West Water, (2012), *Best Practices Guidelines for Water Efficiency*, Sunshine, Australia.

DeOreo WB, Mayer PW. (2012). Insights into declining single-family residential water demands. *J AWWA*. 2012;104(6):E383–E394.

DeOreo, W.B., Mayer, P., (2016), *Residential End Uses of Water, Version 2*, Aquacraft, Inc., Water Research Foundation (WRF), Denver, CO, USA.

Dornier Consulting, 2014, *Residential End Study in Abu-Dhabi, Analysis of Water Use*, Dornier Consultant GmbH (Abu Dhabi), UAE.

Dworak, T., Berglund, M., Laaser, C., (2007), *EU Water Saving Potential*, Ecologic- Institute for International and European Environmental Policy, Berlin.

Fielding, K.S., Russell, S., Spinks, A., Mankad, A., (2012), Determinants of household water conservation: The role of demographic, infrastructure, behavior, and psychosocial variables. *Water Resources Research*, 48, W10510.

Gleick, P.H.P., Haasz, D., Henges-Jecket, C., (2003), *Waste Not, Want Not: The Potential for Urban Water Conservation in California*, Pacific Institute for Studies in Development, Environment, and Security, Oakland, CA, USA.

Goleta Water District, (2018), *Smart Landscape Rebate Program; Single-Family Residential Application*, Goleta Water District, Goleta, CA, USA.

Griggs, J., (2009), *Water Efficiency in New Homes; an Introductory Guide for Housebuilders*, NHBC Foundation, Amersham, UK.

Hanak, E., Davis, M., (2006), Lawns and Water Demand in California, *California Economic Policy 2, No. 2.*, 1–22, USA.

Harrison, M., (2010), Flush: Examining the efficacy of water conservation in dual flush toilets. *SOLAR 2010 Conference Proceedings*, American Solar Energy Society, Phoenix, AZ, USA.

Heberger, M., (2014), Urban Water Conservation and Efficiency Potential in California, *Natural Resources Defense Council*, New York, USA.

Higginson, E., (2009), *Displacement Bag Testing Report*, Veritec Consulting Inc., Mississauga, Canada.

Hussien, W.A., Memon, F.A., Savic, D.A., (2016), *Water Resources Management*, 30, 2931. doi:10.1007/s11269-016-1314-x.

Jorge, C.N., (2014), Efficiency assessment of the household water uses, MSc Thesis in Environmental Engineering, Instituto Superior Técnico, Lisbon.

Lallana, C., Krinner, W., (2001), *Sustainable Water Use in Europe*, European Environment Agency, Denmark.

Lee, M.C., (2014), *Hot Water-Saving via Showers Using Behavior and Water Demand in Taiwan*, National Taichung University of Science and Technology, Taiwan.

Lev, Y., (2012), *Israeli Experience in Water Saving in the Municipal Sector*, State of Israel Water Authority, Tel–Aviv.

Loh, M., Coghlan, P., and Australia, W. (2003). Domestic water use study: In Perth, Western Australia, 1998–2001, Water Corporation.

Maleki-Nasab, A. Abrishamchi, A., and Tajrishy, M., (2007). Assessment of Residential Water Conservation due to Using Low-Flow Fixtures, Sharif University of Technology, Tehran.

Matos, C., Teixeira, C. A., Duarte, A. and Bentes, I. (2013). Domestic water uses: Characterization of daily cycles in the north region of Portugal, *Science of the Total Environment*, 458, 444–450.

Mollenkopf, T., (2015), *Second Independent Review of the WELS Scheme*, Commonwealth of Australia, Australian Government Department of the Environment, 2015, https://www.waterrating.gov.au/about/review-evaluation/2015-review.

Mudgal, S., Benito, P., (2009), *Study on Water Efficiency Standards*, European Commission, DG ENV, Paris.

OECD, (2008), *Water Resources in Agriculture: Outlook and Policy Issues*, OECD Europe.

Office of Energy Efficiency & Renewable Energy, (2018), Landscaping water conservation, U.S. Department of Energy, Accessed: 30.3.2018; www.energy.gov.

Richter, C.P., (2010), *In-House Consumer Study on Dishwashing Habits in four European Countries: Saving Potentials in Households with a Dishwashing Machine*, Schriftenreihe der Haushaltstechnik Bonn, Bonn. Germany.

Sutherin, S., Lombard, K., St. Hilaire, R., (2013), *Southwest Plant Selector: A Mobile App for Homeowners*, The New Mexico Office of the State Engineer (NMOSE), Albuquerque, NM, USA.

The European Water Label, (2015), *The European Water Label Industry Scheme*, European Water Label Scheme Administrator, Water Label Company Limited, Staffs, England.

Tillman, S.M., Bryant, R., (1988), Analysis of the variables affecting water consumption for the city of Hays, Kansas. *Proceedings of the Symposium on Water-Use Data for Water Resources Management, American Water Resources Association, August 1988*, Tucson, Ariz., USA.

US EPA WaterSense, (2015), *Cobb County Showers KSU with Campus-Wide Savings*, Washington, DC.

Whitman, C.T., (2002), *Cases in Water Conservation, How Efficiency Programs Help Water Utilities Save Water and Avoid Costs*, Washington, DC.

Willis, R.M., Stewart, A.S., Giurco, D.P., Talebpour, M.R., Mousavinejad, A., (2013), End-use water consumption in households: impact of socio-demographic factors and efficient devices. *Journal of Cleaner Production*, 50, 107–115.

4 Commercial, Institutional, and Industrial Water Use

Abbas Yari and Saeid Eslamian

Office buildings, hotels, schools, colleges and vocational institutions, hospitals and laboratories, restaurants and other commercial kitchens, laundries, pools, and government and military institutions are major urban commercial/institutional water consumers

Best management practices for commercial, institutional, and industrial facilities promote water-efficient techniques that can be applied across a wide range of facilities with varying water needs

Common provisions to conserve water in different commercial/institutional sectors are appropriate technologies such as high-efficiency toilets requiring not more than 1.3 gallons per flush (4.9 liters per flush), waterless urinals or urinals that flush with 1 gallon (3.8 liters per flush) or less, automatically timed flushing systems, self-closing faucets with flows of 0.5 gallons per min (1.9 liters per minute) for hand-washing, and, if available, the use of non-potable water for flushing.

Significant reductions in water and energy use can also be realized through the wise planning of heating and cooling systems, including cooling towers, boilers, single-pass cooling systems, chilled water systems, boilers, and steam systems.

Cooling towers often represent the largest use of water in institutional and commercial applications, comprising 20–50% or more of a facility's total water use. However, facilities can save significant amounts of water by optimizing the operation and maintenance of cooling tower systems.

4.1 GENERAL

Water resources are a major production factor for most economic sectors in the world. Residents, manufacturing plants, agriculture farms, and tourism rely on a reliable supply of water, often of a pre-determined quality.

This chapter describes a range of cost-effective water-saving devices and practices – some with payback periods of only a few days. The chapter highlights the typical water savings that can be achieved in commercial, institutional, and industrial applications, and explains how to identify the most appropriate devices and practices for specific equipment, processes, or sites.

4.2 COMMERCIAL AND INSTITUTIONAL BUILDINGS

In developed countries, commercial and institutional buildings use a significant portion of municipally supplied water. With so many businesses making up the commercial and institutional sectors, this represents a great opportunity to conserve water. Thus, best management practices for commercial, institutional, and industrial facilities promote water-efficient techniques that can be applied across a wide range of facilities with varying water needs.

Water use by sectors varies widely between each city and territory. Nonresidential water-use sectors differ in how they use water for day-to-day operations. Commercial use consists of the water used by warehouses, stores and shopping centers, restaurants, hotels and related activities, cinemas, offices, and educational and entertainment facilities, and health establishments. Industrial water demand is focused on cooling, processing and manufacturing operations, power generation, sewerage, cleanup, sanitation, and fire protection. Finally, recreational and environmental uses relate to all nonresidential end users that have a value derived from utility provision directly to the consumer. In terms of the small number of empirical studies concerned with industrial and commercial water demand estimation, there are two main findings. First, the price elasticity of demand for commercial and industrial water is substantially higher than residential uses. This suggests that the commercial and industrial demand for water is potentially more price responsive and may therefore indicate opportunities for substitutability between different qualities of water, including recycling. Second, the output elasticity of both industrial and commercial uses is close to elastic, suggesting that a substantial factor accounting for increases in water usage is the growth of output, with water demand increasing proportionately with the output (Worthington 2010).

To ensure strategies, optimize water, and minimize costs, evaluating water use and setting goals are essential. Identifying opportunities to improve the water use efficiency of processes usually involves the deployment of different water management strategies such as water audits, process integration, and the use of advanced water treatment technologies. Water management strategies provide useful insights into any possible process changes that may lead to an increase in water use efficiency and eventually water savings. A water audit is carried out to measure the quantity and quality of water inputs and outputs within a defined boundary, consisting of a single process or set of processes assumed to be operating at a steady state. One of the most useful outcomes of a water audit is the creation of a water flow diagram – an easy to understand representation of usually complex process systems. A water flow diagram offers an idea of how much water is being used by each process, including the volume and quality of the wastewater being generated (Aganda et al. 2013).

4.2.1 Common Measures to Conserve Water in Commercial, Institutional, and Industrial Sectors

While the equipment and processes vary widely, there are opportunities in all commercial and institutional buildings to achieve significant water savings indoors and

TABLE 4.1
Examples of Potential End-Users of Water in Commercial, Institutional, and Industrial Facilities (US EPA WaterSense 2009)

Indoor/Domestic Water	
• Kitchens, cafeterias, staffrooms	• Laundry
• Faucets	• Washing machines
• Distilled/drinking water	• Sanitation
• Dishwashing machines	• Facility cleaning
• Ice machines	• Sterilizers/autoclaves
• Garbage Disposal	• Equipment washing
• Food preparation	• Dust control
• Frozen yogurt and ice cream machines	• Container washing
• Restrooms and showers	• Process
• Faucets	• Photographic and X-ray processing
• Toilets and urinals	
Cooling and Heating	**Outdoor Water Use**
• Cooling towers	• Irrigation
• Evaporative coolers	• Pools and spas
• Boilers and steam systems	• Decorative water feature
• Once-through cooling	
• Air conditioners	
• Air compressors	
• Hydraulic equipment	
• Degreasers	
• Rectifiers	
• Vacuum pumps	

outdoors by making improvements in several operational areas. Despite the differences between subsectors and the factors contributing to their water needs, most corporations have similar water end users, as shown in Table 4.1; (US EPA- WaterSense 2009). For example, domestic water use for plumbing fixtures such as toilets, faucets, showerheads, and urinals represents 25–50% of all water use within most of these facilities. Many of these facilities also utilize a significant portion of their water for irrigation and landscaping. Finally, at least half of these facilities use a considerable amount of their water for heating and cooling purposes. Since domestic water users and landscaping in commercial, institutional, and industrial sectors are also likely in the residential sector; related conservation devices and methods are described in detail in Chapter 3.

Common provisions to conserve water in different commercial/institutional sectors are appropriate technologies such as high-efficiency toilets requiring not more than 1.3 gallons per flush (4.9 liters per flush), waterless urinals or urinals which flush with 1 gallon (3.8 liters per flush) or less, automatically timed flushing systems, self-closing faucets with flows of 0.5 gallons per min (1.9 liters per minute) for handwashing and, if available, and where codes and health departments permit, the use of non-potable water for flushing.

Businesses can save water in landscaping by using the principles of Xeriscape™ (discussed in detail in Chapter 3). Xeriscape is a trademark of Denver Water., an efficiency-oriented approach to landscaping that encompasses seven essential principles that refer to outdoor water use. Artificial grass is also a good alternative for saving water in landscaping.

Submetering helps to ensure that the costs of using water and, if applicable, wastewater disposal are carefully considered. A reflection on actual water use and costs often leads to a credible motivation for the efficiency improvement of water use. Significant reductions in water and energy use can also be realized through the wise planning of heating and cooling systems, including cooling towers, boilers, single-pass cooling systems, chilled water systems, and boiler and steam systems.

4.2.2 Sanitary Water Use

Toilets, faucets, and, to some extent, urinals are found in all commercial and institutional facility restrooms. Showerheads are likely to be found in healthcare facilities, hotels, schools, universities, and gyms, as well as in office buildings and other areas of employment that provide showers for employee use. Laundry equipment, though less common, is found in dedicated laundromats and within hotels and healthcare facilities. Water use attributed to sanitary fixtures and equipment in commercial subsectors is shown in Figure 4.1, which also demonstrates how restrooms are major water users and should be considered more (WaterSense 2012c).

As shown in Figure 4.2, 52% of all commercial, institutional, and industrial consumption in Dallas belongs to toilets; urinals have their highest water use in offices, at 29% (Blackburn 2017).

Potential water savings from the replacement of urinals in commercial buildings in Sofia, Bulgaria, is shown in Table 4.2. The commercial urinals section of the table

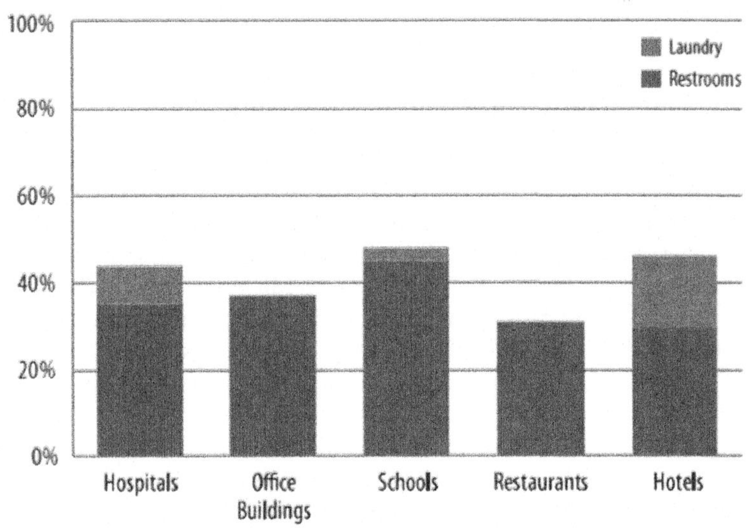

FIGURE 4.1 Water use attributed to sanitary fixtures and equipment (WaterSense 2012c).

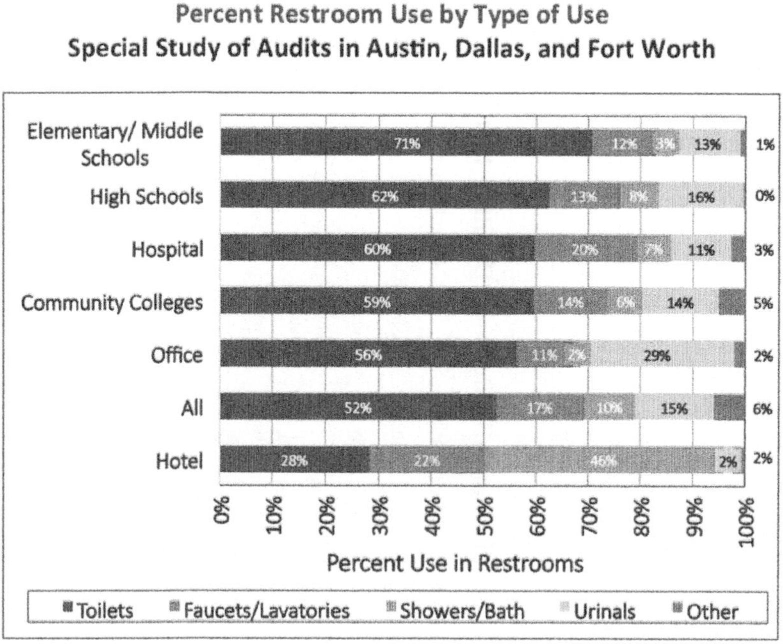

FIGURE 4.2 Restroom water use by type of use (Blackburn et al. 2017).

also shows the result of different control mechanisms in urinal flush. Among all commercial, institutional, and industrial sectors, the sensor-type urinal in schools has the highest potential saving at 91.3% (Dimitrov 1998).

4.2.3 Heating and Cooling

Mechanical systems are used in nearly every type of commercial and institutional facility to provide heating and cooling to buildings. Some facilities also use mechanical methods to cool specific pieces of equipment, such as vacuum pumps, X-ray equipment, and ice machines. In many instances, these mechanical systems use water as the heat transfer medium. As a result, the use of water for building and equipment heating and cooling can be significant, in some cases as much as 30% of the total water use within a facility, as Figure 4.3 shows for various commercial facility types.

Common mechanical systems that use water as the heat transfer medium include single-pass cooling, cooling towers, chilled water systems, and boiler and steam systems.

When looking to reduce mechanical system water use, facilities should first eliminate single-pass cooling, or reuse that water, and then evaluate other cooling and heating systems to maximize efficiency. Single-pass cooling systems use water to remove heat, and cool specific pieces of equipment. However, after the water is passed through the equipment, it is typically discharged to the sewer, rather than being re-cooled and recirculated. In some cases, single-pass cooling can be the single

TABLE 4.2
Potential Water Savings from the Replacement of Urinals in Commercial Buildings in Sofia, Bulgaria (Dimitrov 1998)

Kind of		Urinals (no. of flushes per month)			Urinals (Volume used per month)			Potential saving %		
building	Occupancy	Self-flushing	Cycle	Sendor	Self-flushing	Cycle	Sendor	Self-flushing	Cycle	Sendor
Offices	12	5400	148.5	6534	162	44.55	19.6	Standard	72.5%	87.9%
Restaurants	259	8640	3000	15000	259.2	90	45	Standard	65.3%	82.6%
School	100	4320	650	3750	129.6	19.5	11.25	Standard	85.0%	91.3%
Public WC	1200	14400	9360	36000	432	280.8	108	Standard	35.0%	75.0%

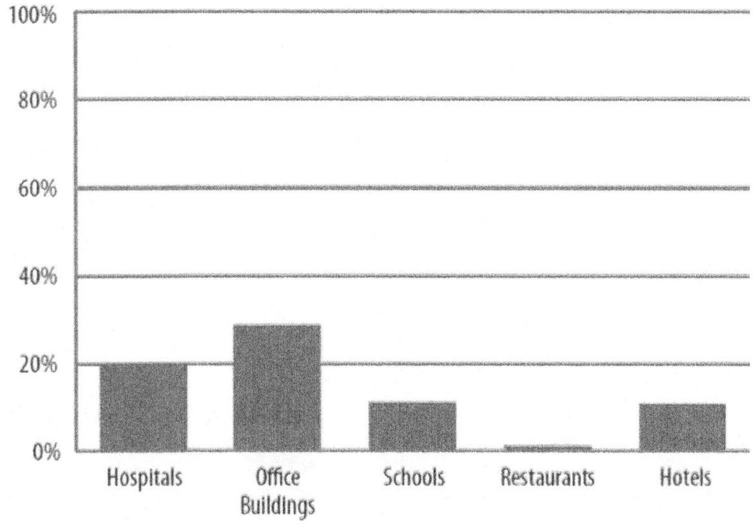

FIGURE 4.3 Water use attributed to mechanical equipment for heating and cooling in various commercial facility types (US EPA WaterSense 2012).

largest water user at a facility, using approximately 40 times more water to remove the same heat load than a cooling tower operating at five cycles of concentration.

Cooling towers are used in a variety of commercial and institutional applications to remove excess heat. They serve facilities of all sizes, such as office buildings, schools, and supermarkets, and larger facilities, such as hospitals, office complexes, and university campuses. Cooling towers dissipate heat from the recirculating water that is used to cool chillers, air conditioning equipment, or other process equipment. By design, they use significant amounts of water. Figure 4.4 shows an evaporative open circuit cooling tower in operation.

Cooling towers often represent the largest use of water in institutional and commercial applications, comprising 20–50% or more of a facility's total water use. However, facilities can save significant amounts of water by optimizing the operation and maintenance of cooling tower systems.

Significant water savings can be achieved by improving the cooling tower management approach. A key mechanism to reduce water use is to maximize the cycles of concentration. Figure 4.5 illustrates this by showing how increasing cycles of concentration can decrease water use in a 100-ton cooling tower. Each facility should determine the maximum periods of concentration it can achieve depending on the quality of the make-up water supply and other facility-specific characteristics (US EPA WaterSense 2012c).

4.3 COMMERCIAL AND INSTITUTIONAL WATER USE

The commercial and institutional sectors are the second-largest consumers of publicly supplied water in the United States, accounting for 17% of the withdrawals from public water supplies (US EPA WaterSense 2012c- Figure 4.6).

FIGURE 4.4 An evaporative open circuit cooling tower in operation.

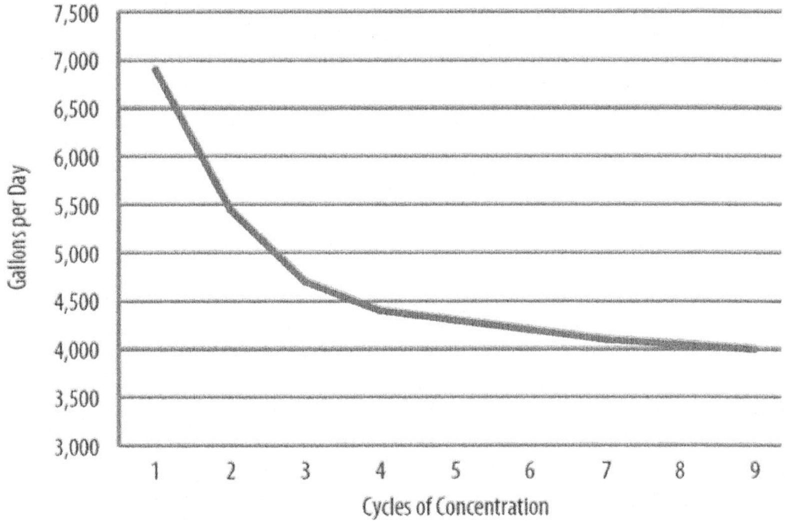

FIGURE 4.5 Cooling tower water use at various cycles of concentration for a 100-ton tower (US EPA WaterSense 2012).

The commercial and institutional sectors include a variety of subsectors, such as hotels, restaurants, office buildings, schools, hospitals, laboratories, and government and military institutions. In Figure 4.7, water used by commercial buildings in Sydney is shown. Manufacturing, commercial, and hospitality have the first, second and, third highest water consumptions, respectively (City West Water 2012). The

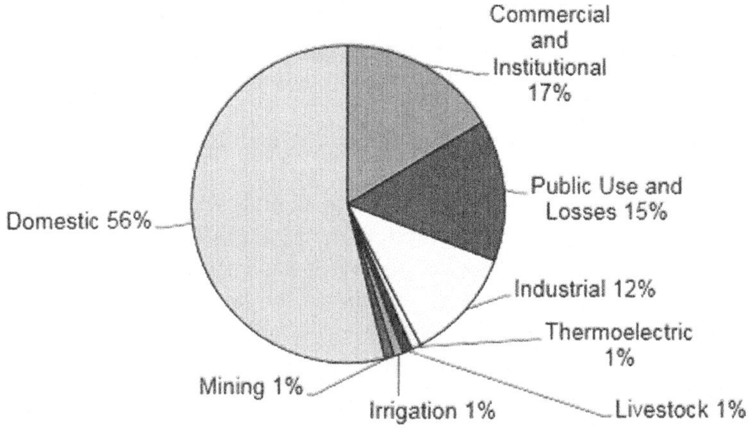

FIGURE 4.6 Urban water use by sectors (US EPA WaterSense 2012c).

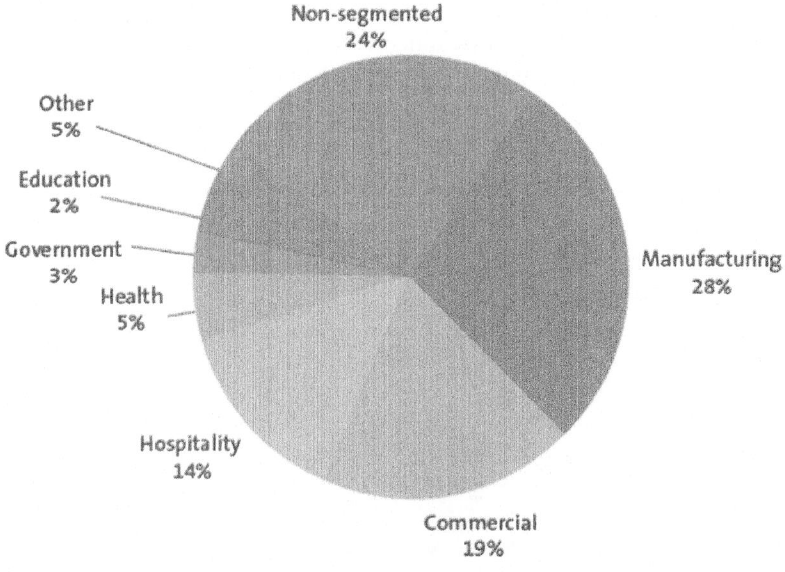

FIGURE 4.7 Water used by commercial buildings in Sydney (City WestWater 2012).

hospitality industry refers to everything from short-term sofa bed lets to star-rated hotels and restaurant establishments, including restaurants/bars, overnight accommodations, and other group shelters.

4.3.1 Distribution of Commercial and Institutional Water Use by Subsectors

Within the commercial and institutional sectors, water use varies by customers (or "customer types") that can be grouped into subsectors. Commercial and institutional water use varies from region to region, due to climate and economic factors that affect the amount of seasonal water use (e.g. landscape water uses and cooling needs in warmer months). Commercial and institutional water use can even vary among water utilities in the same region, depending on the major commercial and institutional customers and the end users of water in each service area.

To evaluate water use by subsectors and to identify which ones typically demonstrate the highest levels of consumption, the EPA analyzed data available from three primary sources in the US on the percent of water use by subsectors. Figure 4.8 displays data compiled from all three sources for subsectors where substantial parity exists between subsector definitions. Despite some variation, all available studies in the US indicate that office buildings, schools, and hospitality and healthcare facilities are likely to be the largest water users when looking at a national breakout. The data presented in Figure 4.8 also indicates that hospitality (restaurants and overnight lodging), office buildings, healthcare facilities, and educational facilities are likely the largest water users in the commercial and institutional subsectors. These results represent the largest U.S. national data sample to date and are consistent with other available studies regarding the subsector's water use within the commercial and institutional subsectors (WaterSense 2009).

4.3.2 End Water Use Pattern in Commercial and Institutional Subsectors

Each facility type has different water use patterns depending upon its function and use. Figure 4.9 shows how water is used in several types of commercial and institutional facilities (US EPA- WaterSense 2012c).

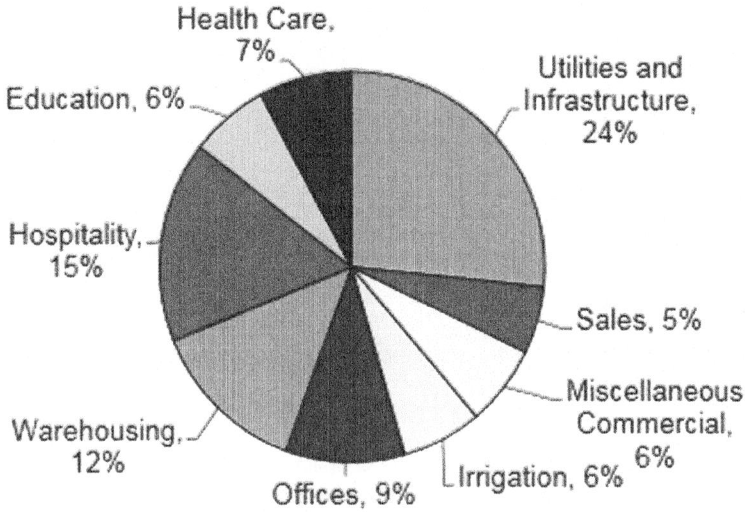

FIGURE 4.8 Estimated distribution of commercial and institutional water use in the United States in 1995 by subsectors (WaterSense 2009).

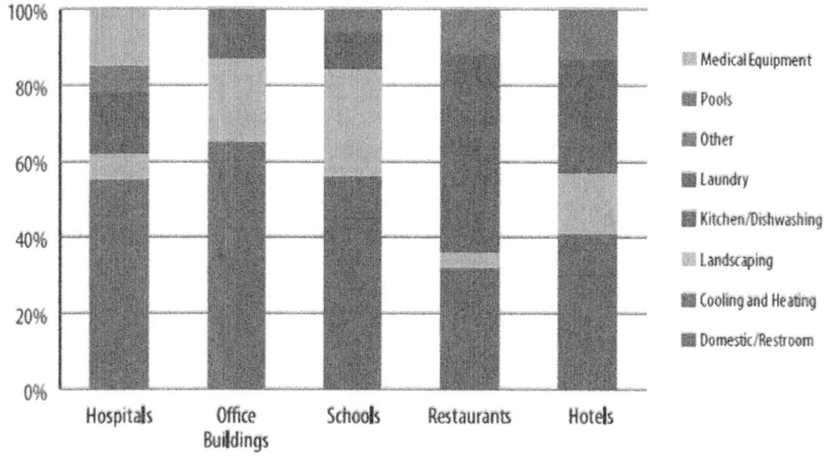

FIGURE 4.9 End water use in various types of commercial and institutional facilities (US EPA WaterSense 2012).

Water is used in different subsectors with distinct characteristics and many differences. In Figure 4.10, the diversity of consumption is shown in Austin, Dallas (Blackburn 2017).

There are plenty of benefits to using water conservation devices in commercial and institutional facilities. Table 4.3 shows a summary of water-saving potential in the public infrastructures of Loire Bretagne basin, France (Dworak 2007).

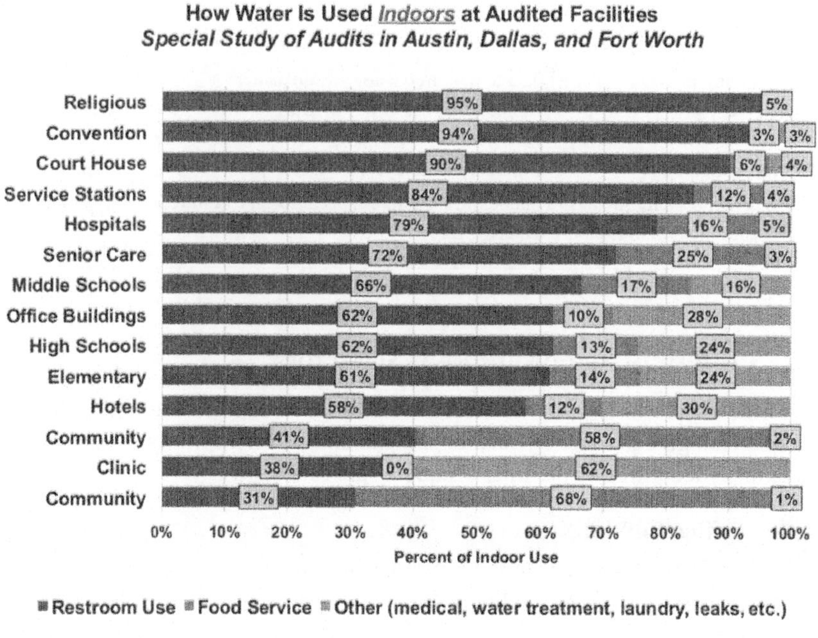

FIGURE 4.10 The diversity of consumption in different sectors (Blackburn et al. 2017).

TABLE 4.3
Public Infrastructures of Loire Bretagne Basin (Dworak et al. 2007)

Water saving potentials in public infrastructures of Loire Bretagne basin - France

The Loire Bretagne water agency conducted a study with the objectives to estimate current water consumptions and potential water savings in various types of public infrastructures: educative buildings, sport equipments, hospitals, administrative buildings, public gardens. Main results are summarised in the following table. Sources of the data are feedback from municipalities experiences, experts estimations, scientific papers or surveys.

Table 25: Saving potential of public buildings in the Loire Bretagne basin

	Consumption oí reference	Savings potential	Source
Primary school	3m³ / child/year	20%	Lonent. Pontivy. Brest. Douamenez. Lannion. Perros. Guirrec
College	General 3,6 m³/student/ year Professional: 6,1m³ / student/year	18%	Conseil régional de Bretagne
Student housing	46,7 m³/ bed/year	30%	CROUS Aquitaine. Eco-Campus
Stadium (normal size)	1000m³/year for equipment use 2000m³/year for irrigation	20%	Surveys CNFPT Midi Pyrénées 2002. AIRES 1998. Report L Cathala
Gymnasium (normal size)	800 m³/an	15%	
Public swimming-pools	0,33 to 0,42 m³/visitor	no data	
Hospitals	100 m³/bed/year	0%	Water agency data, experts
Administrative buildings	14,3 m³/ position/year	20%	water agency data

Average potential water savings in public infrastructures appear to vary from 15% to 30% of the current consumption (hospitals excluded).

4.3.3 Commercial and Institutional Subsectors

Urban water networks supply water for a wide range of commercial and institutional facilities. Though each facility has its specific characteristics, it is common to categorize them according to the most frequent subsectors.

4.3.3.1 Office Buildings

Government and commercial office buildings are among the largest water consumers in some countries. Water audits in twelve of these facilities in Jordan show that significant water and money savings can be made by adopting best practices for water use efficiency (Chebaane 2011).

TABLE 4.4
Benchmarks for Water Efficiency in Sydney Office Buildings (Sydney Water 2007)

Benchmark	Offices with Cooling Towers	Offices without Cooling Towers (Extrapolated Figures)
Median market practice with no leaks	1.01 kiloliters/m²/year	0.64 kiloliters/m²/year
Economic best practice (median of implementing water-saving projects with two-year paybacks)	0.84 kiloliters/m²/year	0.47 kiloliters/m²/year
Very well managed building	0.77 kiloliters/m²/year	0.40 kiloliters/m²/year

Water used in office buildings accounts for approximately 9% of the total water used in commercial and institutional facilities in the United States. The three largest applications of water in office buildings are restrooms, heating and cooling, and landscaping (WaterSense 2012b).

Since bathroom plumbing fixtures account for a significant portion of water use in office buildings, it is smart to assess the age and functionality of existing fixtures. EPA's WaterSense claims that high-performing, WaterSense-labeled fixtures that use at least 20% less water than standard models are available. Upgrades or retrofits can save money and often have short payback periods (WaterSense 2012b).

Table 4.4 shows how much water an average office site in Sydney will consume if it is leak-free (Sydney Water 2007).

To conserve landscape water use, an office complex in Plano, Texas, after upgrading its irrigation system, reduced its outdoor water use by about 40%, saving nearly 45.000 m^3 of water in 2009. With these savings, the project paid for itself in less than a year and a half. To improve the efficiency of the system, the complex upgraded to weather-based irrigation controllers, which analyze local weather data and landscape conditions to program watering schedules based on plants' needs. In addition to installing new controllers, the landscape management firm initiated routine maintenance and repairs to the irrigation system: replacing broken sprinkler heads, positioning sprinkler heads to ensure adequate coverage, and installing pressure-regulating nozzles to increase the uniformity of the water applied. Rain and freeze sensors were also installed to prevent watering at unnecessary times (US EPA- WaterSense 2012c).

4.3.3.2 Hotels

With respect to hotel water use, Figure 4.11 presents the share of different components in the total water demand of an average 3-star hotel in France, showing the relative importance of different components of the total water demand, the irrigation of gardens excluded (Dworak 2007).

As demonstrated in Figure 4.12, showers account for more than a quarter of water use in Melbourne hotels, which is higher than the consumption in a hotel's restaurants and bars, cooling towers, and toilets. The new showerheads are expected to reduce water use at the Westin Melbourne by 12.6 million liters per year (City West Water 2012).

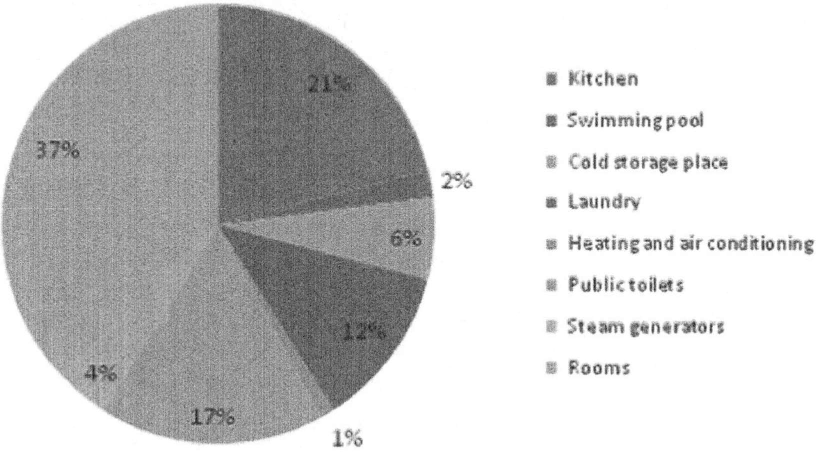

FIGURE 4.11 Different components of a French 3-star hotel, total water demand (Dworak et al. 2007).

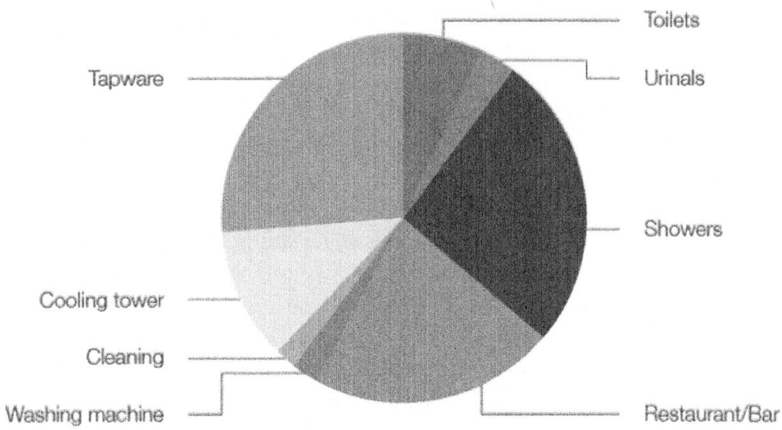

FIGURE 4.12 Water consumption of a hotel with a cooling tower in Melbourne (City WestWater 2012).

In addition to its water-saving initiatives, the hotel has committed to reducing its carbon footprint and is now monitoring energy consumption to help achieve its overall sustainability goals. The showerhead upgrade will make a significant contribution towards these goals, as the energy used for heating water accounts for a significant proportion of the hotel's overall energy consumption. This project alone will reduce gas consumption by 3,600 gigajoules per year and has cut greenhouse gas emissions by 184.9 tons (City West Water 2012).

4.3.3.3 Schools, Colleges, and Vocational Institutions

Schools and other institutes use water in many ways, including some similar to those of the following industries and processes: hospitality, food service, industrial

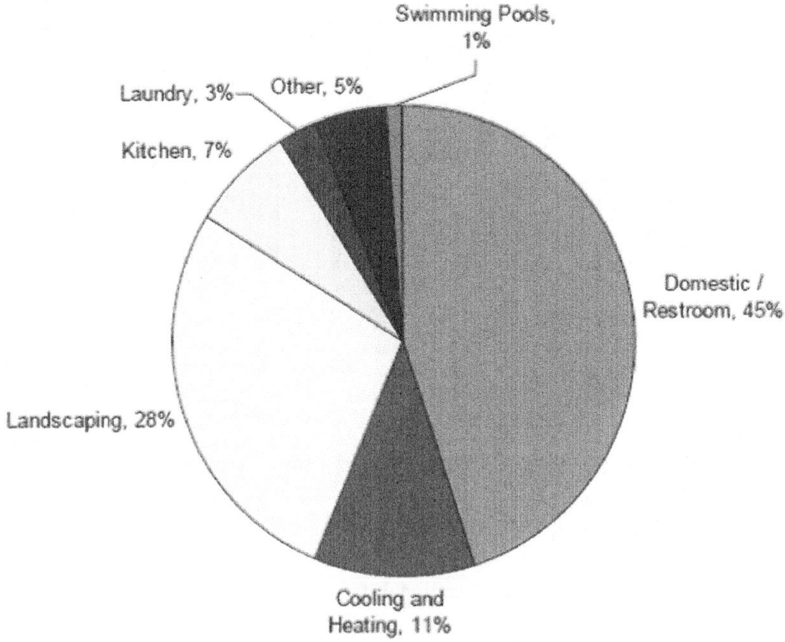

FIGURE 4.13 End users of water in East Bay Municipal Utility District schools, USA (East Bay Municipal Utility District 2008).

laundry, image processing, water purification, vacuum systems, cooling towers and boilers, and cleaning, as well as industrial processes in vocational classes (East Bay Municipal Utility District 2008). The shares of educational water end users' are shown in Figure 4.13.

Six percent of total water use in commercial and institutional sectors in the United States takes place in educational facilities, such as schools, universities, museums, and libraries. The largest users of water in US educational facilities are restrooms, landscaping, heating and cooling, and cafeteria kitchens (WaterSense 2012a; Figure 4.14).

Bathrooms and restrooms are the highest water users in the educational sector, accounting for an average of 28% of the water use in most schools. Repairing leaks in bathrooms and restrooms, and fixing dripping faucets are very efficient solutions for schools. Showerheads, faucets, and toilets that must be replaced due to normal wear-and-tear should be replaced with low-volume models, which are widely available. Low-volume showerheads and faucet aerators can be installed when the entire faucet does not need replacing. An average savings of about 14% of the total water use in schools was possible through this one water conservation action (Southwest Florida Water Management District 2017).

4.3.3.4 Hospitals and Laboratories

From dental and doctor's offices to large general hospitals, veterinary clinics, and research laboratories, medical and laboratory facilities have special operations

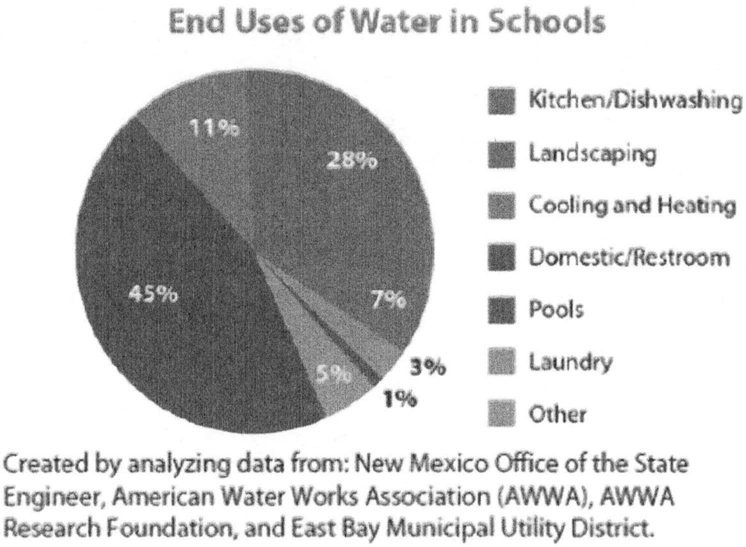

FIGURE 4.14 Water saving in educational facilities in the United States (US EPA WaterSense 2012a).

and equipment. These systems can consume a significant amount of water through water purification, sterilization, photographic and X-ray processes, and vacuum systems. As shown in Figure 4.15, equipment such as steam sterilizers and reverse osmosis systems can account for 5% of a laboratory's total water use (US EPA WaterSense 2012).

Hospitals can attribute more than 15% of their total water use to laboratory and medical equipment, including steam sterilizers and X-ray processing equipment, as shown in Figure 4.16 (US EPA- WaterSense 2012c).

EPA began implementing a comprehensive water management planning strategy for 29 of its laboratory spaces nationwide in the early 2000s and began tracking and reporting water use in 2007. The strategy includes conducting water use and conservation assessments every four years, setting facility-specific, annual water reduction goals, and identifying and implementing water-efficiency projects. When tallied up among all its laboratories, EPA's facility-specific approach to water efficiency has resulted in significant savings. Among all 29 laboratories, water use intensity was reduced by 18.7% between 2007 and 2010, which is equal to a saving of 88.5 million liters of water. Approximately $200,000 in water and sewer costs over the three-year period was also conserved (US EPA- WaterSense 2012c).

4.3.3.5 Restaurants and Other Commercial Kitchens

Restaurants are high water consumers. Water and sewer costs are rising in the U.S. an average of 8% per year, faster than any other utility. Restaurants, which rely on water to efficiently run their kitchens, are feeling the effects of these rising costs. Arby's Restaurant Group, Inc., which is one of the largest restaurant chains in the United States, has worked to combat rising water and energy costs by taking steps to

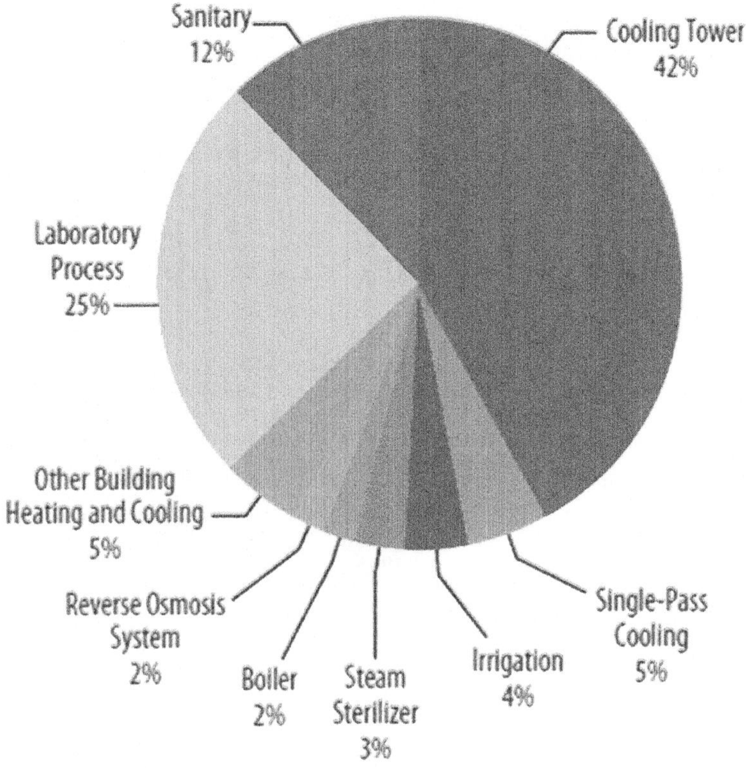

FIGURE 4.15 Laboratory water consumption (US EPA WaterSense 2012).

conserve resources and save money. The chain installed high-efficiency faucet aerators on hand sinks in its restaurants and used vandal-proof units to ensure they would not be stolen (US EPA- WaterSense 2017).

Figure 4.17 is based on the average results from a number of sources. Notably, about 50% of all water use in restaurants goes to kitchens, and 31% to restroom and domestic services (US EPA- WaterSense 2009).

Assessing water-intensive equipment for proper operation and efficiency can help to eliminate water waste. Tools such as dipper wells and wok stoves, for example, can use quite a bit of water because they tend to flow continuously. Additionally, pre-rinse spray valves (fixtures used to remove food particles prior to dishwashing) can have higher flow rates than necessary. Retrofitting or replacing these items with high-efficiency models can be a cost-effective way to reduce water and energy use in commercial kitchens. Table 4.5 provides a summary of the best management practices implemented at each restaurant and an indication of how much water the facilities may be saving compared to typical restaurant practices. The restaurant owners noted that the water- and energy-efficient products and practices have not slowed down productivity in their busy operations, and they are all delighted with the products and equipment they have installed in- and outside the kitchen (US EPA-WaterSense 2012c).

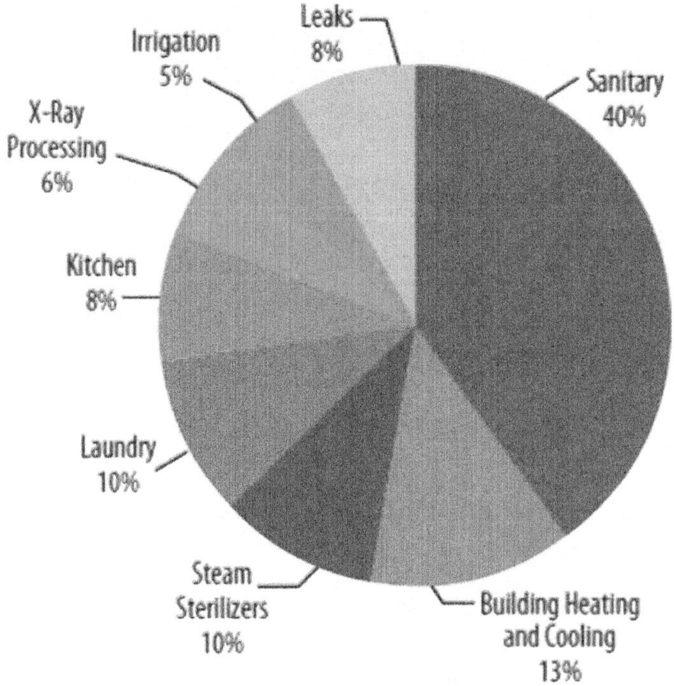

FIGURE 4.16 Hospital water consumption (US EPA WaterSense 2012c).

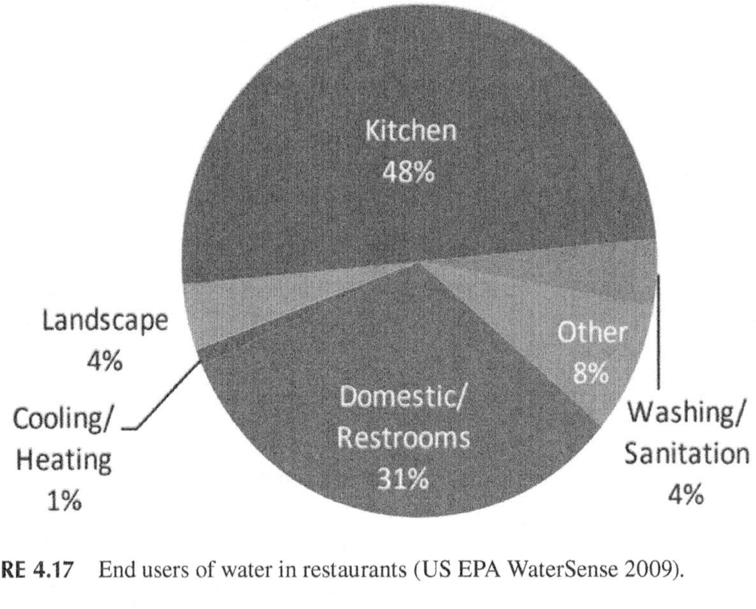

FIGURE 4.17 End users of water in restaurants (US EPA WaterSense 2009).

TABLE 4.5
Best Management Practices Implemented at Certified Green Restaurants; 1 Gallon = 3.785 Liters (US EPA WaterSense 2012-c)

Best Management Practice	Percent Savings Compared to Standard Product/ Practice	Uncommon Ground	The Grey Plume	Founding Farmers
Commercial Kitchen Best Management Practices				
High-Efficiency Faucet Aerators on Prep Sinks	30–75	X	X	X
Manually Operated Dipper Well	Significant			X
High-Efficiency Pre-Rinse Spray Valve	20–40	X	X	X
ENERGYSTAR Qualified Commercial Dishwasher	25	X	X	X
ENERGYSTAR Qualified Commercial Ice Machine	10	X	X	
ENERGYSTAR Qualified Steam Cooker	90			X
Self-Contained Steam Kettle	Significant	X		
Food Composting (no garbage disposal)	100		X	
Other Best Management Practices				
High-Efficiency Aerators on Handwashing Sinks	30–75	X	X	X
Toilets with Dual-flush or 1.28 Gallons per Flush	20	X	X	X
High-Efficiency or Non-Water-Using Urinals	50–100	X		X
Drip Irrigation	20–50	X		
Rainwater Collection and Reuse	Significant	X		

Selecting energy-efficient equipment helps reduce waste heat, which has implications for water use. Because of particular practices in the restaurant and food-service industry, energy-efficient equipment offers significant water savings. Choose refrigerators and freezers that have adequate refrigerator space for thawing food and use air-cooling rather than recalculating cooling water systems. Selecting combination ovens, dry heating tables, and size steam traps for proper operation to avoid dumping condensate, as well as insulating condensate-return lines, using pasta cookers with a simmer mode and automatic over-flow-control valves, using connectionless or boiler-less steamers, and installing in-line restrictors that reduce "dipper well" flow are best practices that offer opportunities for improved water efficiency in food preparation (US EPA- WaterSense 2012c).

4.3.3.6 Laundries

Coin- and card-operated laundries range from those in apartment-complex common rooms to busy commercial laundromats. Laundry operators are installing larger, multi-load machines. The majority of these are hard-mount or solid-mount machines that are bolted to the floor. All multi-load washers can be set to operate at some cycles, including flush, wash, bleach, rinse, scour, and sizing. Also, water levels can be set differently for each cycle, so water use varies greatly depending on the setting. It is important to specify that washers be pre-set to the optimum water factor, which can be done by the factory or by the route operator that leases the equipment.

Currently, the Federal Energy Policy Act Standards of 2005 for commercial coin- and card-operated single-load, soft-mount, and residential-style laundry equipment specify a water factor of 9.5, while the US EPA Energy Star criteria level as of 2007 is set at 8.0. A water factor of 8.0 for all equipment is achievable and recommended (East Bay Municipal Utility District 2008).

The operations of commercial laundries cover a range of applications from on-premises laundries for institutions to commercial activities, such as hotels, nursing homes, hospitals, athletic facilities, and prisons. Industrial laundries offer services for the same set of users as on-premises operations, as well as uniform, diaper, and linen services. Dry-cleaning establishments often have on-premises laundry equipment as well.

4.3.3.7 Pools

Pools and spas are found in many commercial or institutional settings, including hotels, schools, community centers, hospitals, and apartment complexes. Pool covers have been shown to reduce evaporation losses by 30–50%. More efficient filters can reduce water use associated with filter cleaning by 68–98% (US EPA- WaterSense 2012c).

Pools, spas, hot tubs, fountains, and other types of decorative water features can waste large volumes of water if not properly designed and equipped for efficient operation. Four principles govern good practice in this context:

- Design the mechanical equipment to filter, clean, and operate the pool;
- Design the pool to minimize water loss;
- Choose alternatives that use less water;
- Install decorative water features only where they provide tangible benefits.

Pools, spas, and fountains require water for make-up, evaporation, splash-out, filling, backwashing the filter, and replacing water lost to leaks. Evaporation and splash-out vary based upon both weather conditions and activity. In the summer, evaporation alone can amount to 5–12 inches a month, and splash-out can add several inches. This water must be replaced to keep the pool at proper levels (East Bay Municipal Utility District 2008).

4.4 INDUSTRIAL WATER USE

Industrial facilities often use water to move products, change or maintain temperature, clean equipment, or prevent drying between the stations of a manufacturing assembly line. Saving water in the industry can be achieved through a combination of behavior changes, the switching and/or replacement of equipment with water-saving equipment to reduce overall water consumption, and an increase in water reuse. To ensure the success of the strategies, optimize water, and minimize costs, it is important to assess current water use and set goals (Danielsson 2018).

Optimization of water use by industries is important because it can lower water withdrawals from local water sources, thus increasing water availability and improving community relations, increasing productivity per water input, decreasing wastewater discharges and their pollutant load, reducing thermal energy consumption, and potentially processing cost.

Water is distributed in an uneven manner between industries. For example, Figure 4.18 shows industrial water use in Texas in 2010. Reducing water consumption is one of the easiest and most inexpensive ways of achieving cost savings. Many companies can save up to 30% of their water costs by implementing simple and inexpensive water minimization measures (Blackburn et al. 2017).

A cost-effective application is often site-specific. In particular, water-saving devices and practices proposed for industrial processes should be evaluated by those who have a working knowledge of the processes before they are implemented.

The four phases of a type commercial, institutional, and industrial water-saving campaign are as follows:

PHASE 1 – Initiation

- Involve staff and appoint the leader of the water-saving team;
- Find out about and follow up on related water-saving devices and their application;
- Talk to other interested people in the company;
- Develop a simple program;
- Allocate sufficient resources.

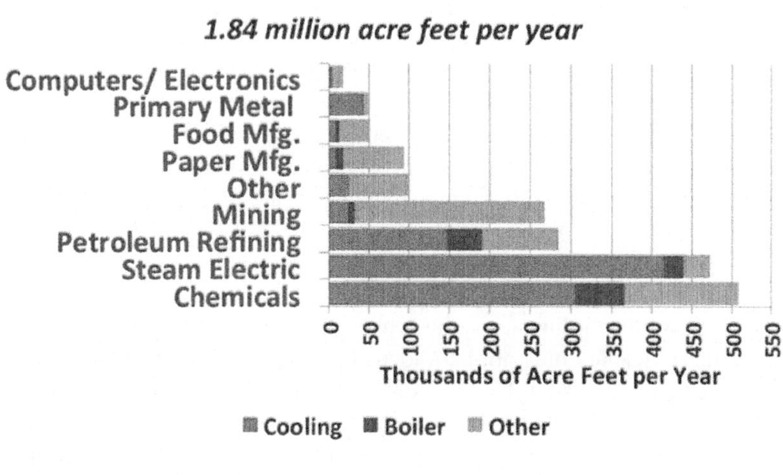

FIGURE 4.18 Industrial water use in Texas, 2010 (Blackburn et al. 2017).

PHASE 2 – Water use survey and development of the water balance

- Identify where, how, and why water is used;
- Identify the water quantity and quality requirement at each point of use;
- Determine the water quality and availability at each point of discharge.

PHASE 3 – Evaluation of water-saving options

- Evaluate current and future water costs by area or by item of equipment;
- Identify and evaluate cost-effective water-saving devices and practices;
- Carry out trials of possible options.

PHASE 4 – Implementation

- Train staff (if necessary);
- Implement cost-effective water-saving devices and practices;
- Monitor the implemented methods and practices;
- Obtain feedback from staff;
- Communicate successes and savings to employees (WRAP- Ripple effect 2005).

Many companies can achieve a 20–50% decrease in the amount of water used in implementing water minimization measures. There is a range of cost-effective water-saving devices and practices, some with payback periods of only a few days. Many guidelines, such as Envirowise (relating to water efficiency for industry and business), highlight the typical water savings that can be achieved for commercial and industrial applications and explain how to identify the most appropriate devices and practices for specific equipment, processes, or sites (WRAP- Ripple effect 2005).

Water and effluent charges per unit of production or service could harbor a considerable portion of final product costs, especially in arid areas, insofar as some manufacturers are broken due to lack of water.

Considering the use of submetering to identify the amount of water used by different users, such as landscaped areas, cafeterias, laundries, and major industrial equipment is important. Metering all water use in the system, also known as universal metering, provides critical data for water system management and planning purposes. Universal metering of both public and private accounts is a water industry best practice (American Water Works Association 2013). Also, this data provides important information about water end use that supports more targeted water conservation and efficiency programs and policies.

Price signals may influence some water use, particularly for peaking water use, but water efficiency may not be a focus for many others. Incentivizing site- or customer-specific audits may be the most effective means of identifying opportunities for achieving an efficient implementation for commercial, institutional, and industrial customers. In many situations, there may be considerable overlap with residential efficiency considerations (e.g. faucets, showerheads, toilets, and outdoor irrigation). In others, however, particular consideration for the needs of the facility may focus on significant opportunities for water and financial savings (e.g. healthcare, food prep). A range of WaterSense and ENERGY STAR labeled products may be appropriate for use in commercial, institutional, and industrial facilities.

WaterSense has a comprehensive best management practices guide that identifies additional opportunities for savings. In some cases, it may be possible to manage peak demands (e.g. with incentives to industrial customers or for behavior change in residential customers) to even out demand and reduce withdrawal and/or storage needs. Accurate information on water use trends is critical to understanding how well a system is managed and where improvement can be sought (US EPA 2016).

The California Urban Water Conservation Council has identified the replacement of liquid ring vacuum pumps with dry vacuum systems as a potential best management practice (PBMP). This investigation was undertaken to assess the potential viability of dry vacuum pump systems as a best management practice (BMP) for urban water purveyors in California. There is significant potential for cost-effective water savings benefits from the adoption of dry vacuum pump technology. This conclusion may be even stronger when wastewater and energy impacts are also included in the benefits analysis. Some California water agencies already have rebate programs for dry vacuum pumps or mechanisms such as customized rebate programs for commercial, institutional, and industrial customers that can handle rebates for dry vacuum systems. Vacuum pumps and systems are presently used in a wide range of commercial and industrial sectors including dental, medical, food handling, processing, packaging, pharmaceutical, chemical, electronics manufacturing, light bulb manufacturing, scientific/laboratory, petroleum, groundwater remediation, pulp, and paper mills and sewage treatment (Fryer 2012).

4.5 CONCLUSION

As drought events and water scarcity situations are becoming more common, there is an urgent need for policy action to tackle issues surrounding water use, and to ensure a clear, sustainable future for water resources and management. Consumers can choose from a wide range of technical possibilities to save water without changing their systems or behavior drastically.

Water-saving potentials differ between sectors and different regions, but many companies can achieve a 20–50% decrease in the amount of water they use by implementing water minimization measures.

REFERENCES

Agana, B. A. Reeveb, D., Orbellab, J. D., 2013, An approach to industrial water conservation–A case study involving two large manufacturing companies based in Australia. *Journal of Environmental Management*, 114(15), 445–460.

American Water Works Association, 2013, *AWWA Standard G480-13*, Water Conservation Program Operation and Management, Denver, CO, USA.

AS/NZS 6400, 2016, *Water Efficient Products: Rating and Labeling*, Australian/New Zealand Standard 6400.

Blackburn, J., Hoffman, H. W., Clements, D., Water, B., 2017, *Pace a New Tool for Big Water Savings in the CII /Sectors*, Texas Living Waters Project, Austin, TX, USA.

Chebaane, M., Hoffman, B., 2011, *Office Buildings Water Efficiency Guide, Water Demand Management Program*, U.S. Agency for International Development, Washington, D.C..

City West Water, 2012, *Best Practices Guidelines for Water Efficiency*, Business Resource Efficiency, Victoria, Australia.

Danielsson, M., Spuhler, D., 2018, Reduce water consumption in industry. Sustainable sanitation and water management toolbox, SSWM.info. https://www.sswm.info/using-sswm-content [June 2018].

Dimitrov, G., 1998, *Technical Possibilities for Reducing the Wastage of Water in Residential Buildings Supply Networks*, University of Architecture, Civil Engineering and Geodezy, Sofia.

Dworak, T., Berglund, M., Laaser, D., Strosser, P., Roussard, J., Grandmougin, B., Kossida, M., Kyriazopoulou, I., Berbel, J., Kolberg, S., Rodríguez-Díaz, J. A., Montesinos, P., 2007, *EU Water Saving Potential*, Ecologic-Institute for International and European Environmental Policy, Part 1 –Report, Berlin.

East Bay Municipal Utility District, 2008, *A Water-Use Efficiency Plan Review Guide for New Businesses, Watersmart Guidebook*, EBMUD., San Francisco.

Fryer, J., 2012, *Water Use in Vacuum Pump Systems & Viability for a Water Conservation Best Management Practice in California*, California.

Kelly, D. A., 2013, The European Water Label: An analysis and review. *CIBW062 Symposium: 39th International Symposium on Water Supply and Drainage for Buildings*, Nagano, Japan.

Southwest Florida Water Management District, 2017, *School Checklist*, Water Conservation Program, Brooksville, FL, USA.

Sydney Water Corporation, 2007, *Best Practice Guidelines for Water Conservation in Commercial Office Buildings and Shopping Centers*, Sydney.

The E, uropean Water Label, 2015, The European water label industry scheme.

US EPA WaterSense, 2009, *Water Efficiency in the Commercial and Institutional Sector*, United States Environment Protection Agency, Washington DC.

US EPA WaterSense, 2012, *WaterSense at Work: Best Management Practices for Commercial and Institutional Facilities, Water Efficiency in the Commercial and Institutional Sector*, United States *for Water Supply Expansion*, Washington, DC.

US EPA WaterSense, 2016, *Best Practices to Consider When Evaluating Water Conservation and Efficiency as an Alternative for Water Supply Expansion*, Washington DC.

US EPA WaterSense, 2017, *Major Restaurant Chain Slows the Flow with a Water-Saving Strategy for Success*, United States Environment Protection Agency, Washington DC.

Worthington, A. C., 2010, *Commercial and Industrial Water Demand Estimation: Theoretical and Methodological Guidelines for Applied Economics Research, Department of Accounting*, Finance and Economics, Griffith University, Mount Gravatt, Australia.

WRAP Ripple Effect, 2005, *Cost-effective Water-Saving Devices and Practices for Industrial Sites*, Envirowise Publication, Oxon, England.

5 Water Efficiency Regulations in Domestic Buildings

Abbas Yari and Saeid Eslamian

Depending on the culture of water use and the specific territorial resources available, some countries have introduced laws and regulations to restrict water use through the labeling of appliances according to water use. It is expected that these standards will control the production and import of water-consuming devices, and create grounds on which to force the use of standard appliances by aiding in the supply, distribution, and installation of these water-saving devices in public buildings, local authorities, and newly constructed buildings.

Appliance labeling is a powerful educational tool. The success of the Energy Star labeling program highlights the power of information. A "Water Star" label for water-using appliances should be implemented, showing total water use per year (or some comparable measure). Such labeling permits consumers to make more informed choices about their actions and purchases.

Major global regulations are the WaterSense Program in the United States, the Australian Water Efficiency Labeling Scheme (WELS), the Waterwise Project in the UK, the European Union's Water Label, the Water Efficiency Labelling Scheme in New Zealand, the Water Efficiency Labelling Scheme in Singapore, and China's Water Conservation Certificate.

5.1 GENERAL

Consumers have little information about the water-saving capabilities of water conservation devices and methods, and even less about the environment and the financial benefits that water-efficient devices can offer. As a result, depending on the culture of water use and the specific territorial resources available, some countries have introduced laws and regulations to restrict water use through the labeling of appliances according to water use. It is expected that these standards will control the production and import of water-consuming devices, and create grounds on which to force the use of standard appliances by aiding in the supply, distribution, and installation of these water-saving devices in public buildings, local authorities, and newly constructed buildings. The grading and labeling of water equipment enables the consumer to choose the most appropriate appliance by comparing productivity. When examined under a Water Efficiency Labelling Scheme, a machine receives higher number grades or more stars by increasing its efficiency. Several rating schemes have been set up worldwide to measure the consumption of water-consuming equipment, and some of the most common labels, from various different countries, are discussed below.

The water efficiency schemes of different countries are in different stages of development and are implemented in several forms. In some countries, it is a compulsory requirement to provide water efficiency labels for certain kinds of plumbing fixtures and appliances before they can be put on sale in the market. For others, the water efficiency schemes are implemented on a voluntary basis so as to allow lead time for the market to move towards more water-efficient products.

Although the primary metric of the scheme is water, it is recognized that energy saving is the key to encourage behavior change and to raise consumer awareness of the importance of using water more wisely. It is widely reported that if consumers can save money they are more likely to "buy-in" to an initiative. A quarter of household energy bills are associated with hot water. With this in mind, industry experts have developed an energy aspect to the labels for taps and showers that provides an annual energy calculation (The European Water Label 2015). The summary of the water efficiency standard label is presented in Figure 5.1 (Kelly 2013).

5.2 US WATERSENSE PROGRAM

WaterSense® is a voluntary partnership program sponsored by the U.S. Environmental Protection Agency (EPA), designed to make it easy for Americans to save water and protect the environment by choosing water-efficient products and services. Items that meet WaterSense specifications must be independently tested and/or certified, and only then can they carry the WaterSense label; Figure 5.2 shows a WaterSense Label.

		Europe	**Portugal**	**Hong Kong**	**Singapore**	**Australia**
Device		*Voluntary*	*Voluntary*	*Voluntary*	*Mandatory*	*Mandatory*
Toilets	*(l/flush)*	$3.5 \geq L > 6.0$	$4.0 \leq L \leq 9.0$		$3.5 \geq L < 4.5$	$2.5 > L < 5.5$
Showers	*(l/min)*	$6.0 \geq Q > 13.0$	$5.0 \geq Q > 30.0$	$9.0 \geq Q > 16.0$	$5.0 \geq Q < 9.0$	$4.5 < Q > 16.0$
Taps	*(l/min)*	$6.0 \geq Q > 13.0$	$2.0 \geq Q > 8.0$[1] $4.0 \geq Q > 10.0$[2]	$2.0 \geq Q > 6.0$[3] $5.0 \geq Q > 9.0$[4]	$2.0 \geq Q > 6.0$[1] $4.0 \geq Q > 8.0$[2]	$4.5 > Q > 16.0$
Urinals	*(l/flush)*	$L = 1.5$		$1.5 \geq L > 4.5$	$0.5 \geq L < 1.5$	$7.0 > L > 2.5$
Baths	*(l/bath)*	$155 \geq L > 200$				
Washing Machine	*(l/kg/cycle)*			$9.0 \geq L > 13.0$[5] $16.0 \geq L > 22.0$[6]	$9.0 \geq L < 15.0$	X
Dishwashers						X
Flow controllers		X				X
Greywater system		X				
Electric showers		X				

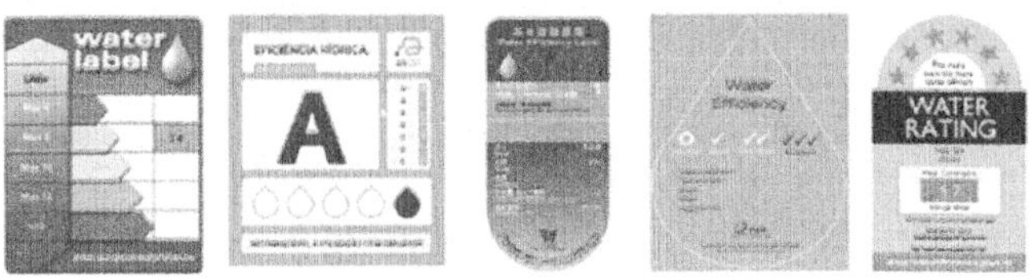

Notes: 1 Bathroom taps; 2 Kitchen taps; 3 Non-mixing taps; 4 Mixing taps; 5 Horizontal drum washing machine; 6 Impellor type washing machine

FIGURE 5.1 Summary of water efficiency standard labels for selected water-using device labeling schemes.

FIGURE 5.2 WaterSense label.

In December 2006, the EPA launched the WaterSense program as part of its Sustainable Infrastructure Program. The mission of the WaterSense program is to "protect the future of the nation's water supply by promoting and enhancing the market for water-efficient products and services". The program also aims to establish public–private partnerships to encourage water conservation among manufacturers, developers, and consumers.

WaterSense was developed to help multifamily housing property owners and managers improve their water management, reduce property water use, and subsequently improve their EPA Water Score. However, many of the best practices in the WaterSense guides can be used by facility managers for non-residential properties.

The WaterSense program is multifaceted in that it:

- Provides information to consumers and businesses on water-efficient, high performing products, homes, and practices;
- Educates consumers on the importance of water efficiency;
- Ensures water-efficient product performance;
- Promotes innovation in product development;
- Holds certification programs for services (e.g. irrigation professionals);
- Offers a water efficiency labeling scheme.

With regards to labeling, manufacturers that receive the label from approved and licensed certifying bodies may only share it with retailers, distributors, and wholesalers for promotional purposes (Mollenkopf 2015).

5.3 AUSTRALIA WATER EFFICIENCY LABELING SCHEME (WELS)

The Australian Water Efficiency Labeling Scheme (WELS) is a labeling scheme for water efficiency. The Australian Standard Terms and Conditions are defined to encourage citizens to select products based on the WELS criteria. In the WELS model, washing machines and dishwashers, as well as equipment in toilets, baths, washrooms, laundries, and kitchens, alongside other valves and flow control equipment, are ranked to help citizens make informed decisions when buying these products.

The objective of the WELS scheme is to promote the use of water-efficient products by enabling consumers to clearly identify and purchase water-efficient products with confidence that the performance of these products has not been compromised due to their reduced water use. It is also intended to encourage manufacturers to develop and market products that are more water-efficient. All products that are required to be labeled under the WELS scheme will be rated. Under the WELS scheme, washing machines, dishwashers, lavatory equipment, showers, tap equipment, and urinal equipment are required to be registered and labeled. Lavatory equipment is also required to comply with minimum water efficiency requirements. Flow controllers may be voluntarily registered and labeled. If a product is voluntarily registered and labeled, it is also required to comply with all the relevant aspects of the standard.

Registration is renewed annually. A test report from an approved laboratory must be provided to prove compliance with the scheme's water efficiency requirements (AS/NZS 6400:2016). There are plans to extend the scheme to include washer dryers, evaporative air conditioners, instantaneous gas hot water systems, hot water recirculators, and irrigation systems (European Commission 2009). Applicants for the label are certified against a set of criteria and products are tested by designated institutions. Once a product has been certified, it is listed by the Department of State Economic and Trade Commission and is given priority in government procurement (Mollenkopf 2015).

5.4 UK WATERWISE PROJECT

Waterwise is an independent, not-for-profit, non-governmental organization focused on decreasing water consumption in the UK and building the evidence base for large-scale water efficiency. The objective was for Waterwise to support water companies in their work to promote water efficiency to their customers. Waterwise seeks to build on actions taken by water companies and to facilitate partnerships that enable large-scale innovative initiatives that encourage water to be used more wisely across the UK. As a result of this work, Waterwise has become the leading authority on water efficiency in the UK and Europe. The purpose of Waterwise is to innovate, develop, and test new ideas, and to help demonstrate practical water efficiency actions. It works in partnership with other organizations and communities to help deliver water efficiency projects. Waterwise acts as a knowledge hub for innovative products and approaches. It aims to incubate, develop, and match innovation and innovators with other partners, including through product and service accreditation schemes (Waterwise 2017).

5.5 THE EUROPEAN UNION (EU) WATER LABEL

The European Union introduced the Water Label as a voluntary label. The Water Label launched in 2014 and aimed to promote water-efficient devices, and to provide

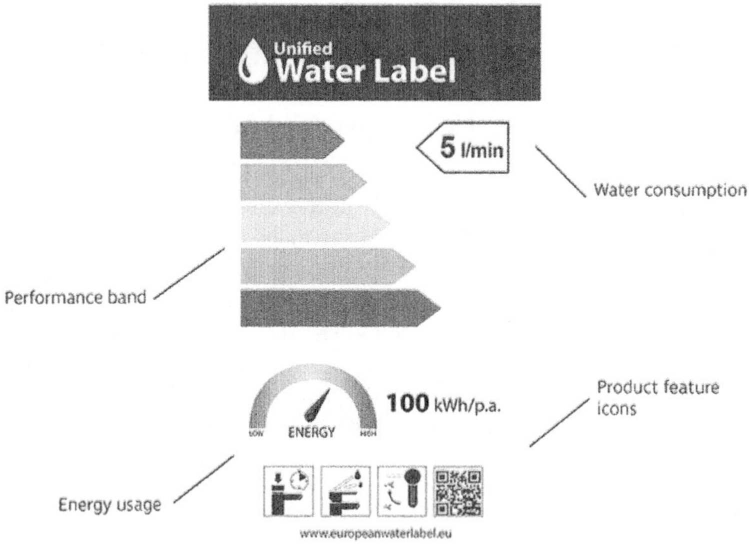

FIGURE 5.3 European Water Label (European Water Label 2018).

consumers with information on the water-consumption and efficiency ratings of water-using devices in order to influence their choice in selecting more water-efficient devices. The Water Label is applicable, not only throughout the EU, but also in Israel, Switzerland, Russia, Ukraine, and Turkey. (Figure 5.3) illustrates the main form of the Water Label (European Water Label 2018). The primary aim of this new labeling scheme is to provide a single classification system across all member countries to inform consumers of the water consumption of water-using devices. The Water Label will cover a wide range of devices including showerheads, shower controls, electric showers, taps, toilets, baths, urinals, greywater systems, and flow regulators (Kelly 2013).

The voluntary scheme currently has 66 major brands registered from businesses across Europe, and a database of registered products that continues to grow. The EWL covers the following bathroom products:

- Toilets;
- Baths;
- Taps – including in kitchens;
- Showers – electrics, handsets, and valves;
- Flow regulators;
- Independent flushing cisterns;
- Urinal controllers;
- Replacement toilet flushing devices;
- Greywater products.

The label shows the amount of water that each product uses (liters per minute). It was designed to be similar to other energy efficiency labels in circulation and to mirror the colors and performance bands found on these labels (Mollenkopf 2015).

5.6 NEW ZEALAND WATER EFFICIENCY LABELLING SCHEME

The New Zealand Water Efficiency Labelling Scheme became mandatory on 1 April 2011. The New Zealand Scheme is based on the same standard as the Australian WELS scheme (AS/NZS 6400:2016): it mirrors much of the administrative arrangements, relies on the Australian Scheme for the registration of products (as there is no such process in New Zealand), and uses the same labels that are used under the Australian Scheme. The following products are required to be registered with the New Zealand Water Efficiency Labelling Scheme: washing machines, dishwashers, lavatories, showers, taps, and urinals (Department of the Environment 2015). The notable exception, when compared to the Australian WELS Scheme, is that of flow controllers.

5.7 SINGAPORE WATER EFFICIENCY LABELLING SCHEME

In October 2006, the Singapore Public Utilities Board (PUB) and the Singapore Environment Council (an independent charity) introduced a voluntary water efficiency labeling scheme in order to provide information to consumers on the water consumption and efficiency of products and fittings.

The voluntary initiative was part of an umbrella program called the "10-litre challenge" aimed at encouraging Singaporeans to reduce their daily water consumption by 10 liters per day. In light of the favorable response to the program, and in an attempt to further encourage the use of water-efficient products and fittings, it was decided to mandate the labeling scheme.

The PUB enacted the Public Utilities (Water Supply) (Amendment) Regulations 2008, establishing the Mandatory Water Efficiency Labelling Scheme (MWELS) and the mandatory installation of dual-flush, low-capacity cisterns. In addition, all new domestic and non-domestic premises, as well as existing premises undergoing renovation, were required to install MWELS products and fittings. The MWELS came into effect on the 1 July 2009.

The following products are required to conform to mandatory standards set by the PUB regarding their supply, sale, and use in Singapore:

- Shower taps and mixers;
- Basin taps and mixers;
- Sink and bib taps and mixers;
- Bath and shower taps and mixers (except for concealed ones);
- Dual-flush, low-capacity flushing cisterns;
- Urinal flush valves and waterless urinals;
- Washing machines intended for household use.

Showerheads remain under a voluntary labeling scheme. Vendors of these products can voluntarily demonstrate that their product meets the PUB's water efficiency labeling scheme standards and requirements (Mollenkopf 2015).

TABLE 5.1
Comparing Various Water Efficiency Schemes (Mudgal et al. 2009)

			Products Covered							
			Indoor Wup							Outdoor Wup
Name of the Scheme	**Geographical Coverage**	**Mandatory/ Voluntary**	**Toilets**	**Washing Machine**	**Dishwasher**	**Showerheads**	**Taps**	**Urinals**	**Car Wash**	**Garden Irrigation Equipment**
Outside the EU										
Australia WELS	National	M	×	×	×	×	×	×	×	
New Zealand Wels	National	M		×	×	×	×	×		
USA Energy Policy Act	National	M	×	×	×	×	×	×	×	
Singapore WELS	National	M	×	×			×	×	×	
Smart Approved WaterMark (Australia)	National	V							×	×
Car Wash Rating Scheme (Australia)	National	V							×	
Hong Kong WELS	National	V					×			
WaterSense (US)	National	V	×			×	×	×		
U.S. Energy Star	National	V		×	×					
Singapore Green Label	National	V		×						
Korea Green Label	National	V	×				×	×		
Thailand Green Label	National	V	×	×			×		×	
Japan EcoMark	National	V	×					×	×	
Within the EU										

(Continued)

TABLE 5.1 (CONTINUED)
Comparing Various Water Efficiency Schemes (Mudgal et al. 2009)

			Products Covered							
			Indoor Wup							Outdoor Wup
Name of the Scheme	Geographical Coverage	Mandatory/ Voluntary	Toilets	Washing Machine	Dishwasher	Showerheads	Taps	Urinals	Car Wash	Garden Irrigation Equipment
Ordenanza de Gestión y Uso Eficienete del Agua (Spain)	Municipal	M	×			×	×		×	
The EU Ecolabel	EU	V		×	×					
Ecodesign Directive (includes possible extension to energy-related products)	EU	M		×	×	×	×			
Distintivo de Garantía de Calidad Ambiental Catalan (Spain)	Regional	V	×			×	×			
Ambientale al Regolamento Edilizio della Città di Avigliana (Italy)	Municipal	M	×			×	×			
Variante all'Art. 8 delle Norme Tecniche di Attuazione del P.R.G (Italy)	Municipal	M	×							
Regolamento Energetico Ambientale (Italy)	Provincial	M	×			×	×			

(Continued)

TABLE 5.1 (CONTINUED)
Comparing Various Water Efficiency Schemes (Mudgal et al. 2009)

			Products Covered							
			Indoor Wup							Outdoor Wup
Name of the Scheme	Geographical Coverage	Mandatory/ Voluntary	Toilets	Washing Machine	Dishwasher	Showerheads	Taps	Urinals	Car Wash	Garden Irrigation Equipment
Certificação da eficiência Hídrica de Produtos (Portugal)	National	V	×	×	×	×	×			
Water Supply (Water Fittings) Regulations (UK)	National	M	×							
BMA Water Efficiency Labelling Scheme (UK)	National	V	×			×	×			
The Blue Angel (Germany)	National	V	×					×		
The Nordic Ecolabel	Transnational	V		×	×				×	

5.8 CHINA WATER CONSERVATION CERTIFICATE

The China Water Conservation Certification (CWCC) is available to a broad range of products (62 different categories in total). Some of these categories include industry (e.g. cooling towers and automatic filters), agriculture (e.g. irrigation equipment), and domestic (e.g. taps and showerheads).

Conformity with the CWCC standards is managed by the China Energy Conservation Product Certification Centre. Certification is voluntary and aims to encourage the innovation of more water-efficient products while ensuring consumers have information to make more sustainable purchase decisions.

5.9 WATER LABELS COMPARISON

Table 5.1 shows the differences between water efficiency schemes throughout the world (Mudgal 2009).

5.10 CONCLUSION

The largest, least expensive, and most environmentally-sound source of water to meet future needs is the water currently being wasted in every sector of the economy. Improved efficiency and increased conservation are the cheapest, easiest, and least destructive ways to meet current and future water needs. Communities could save 30% of their current urban water use with cost-effective water-saving solutions. Indeed, fully implementing existing conservation technologies in the urban sector can eliminate the need for new urban water supplies for years to come. Many technologies for using water more efficiently are available in every urban sector. These include low-flow toilets, faucets, and showerheads, drip and precision irrigation sprinklers, commercial and industrial recycling systems, and many more. Existing technologies for improving urban water conservation and water-use efficiency also have enormous as-yet unexploited potential.

Appliance labeling is a powerful educational tool. The success of the Energy Star labeling program highlights the power of information. A "Water Star" label for water-using appliances should be implemented, showing total water use per year (or some comparable measure). Such labeling permits consumers to make more informed choices about their actions and purchases.

REFERENCES

AS/NZS 6400, 2016, *Water Efficient Products: Rating and Labeling*, Australian/New Zealand Standard Parliament of Western Australia, Canberra, Australia.

Department of the Environment, 2015, New Zealand Water Efficiency Labelling Scheme Label, Parliament of Western Australia, Canberra, Australia.

European Commission, 2009, Study on Water Efficiency Standards, Final Report, DG ENV, Paris.

European Water Label, 2018, One product label across Europe for all water using bathroom products. http://www.europeanwaterlabel.eu/thelabel.asp [June 2018].

Kelly, D. A., 2013, The European Water Label: An analysis and review. *CIBW062 Symposium: 39th International Symposium on Water Supply and Drainage for Buildings*, Nagano, Japan.

Mollenkopf, T., 2015, *Second Independent Review of the WELS Scheme*, Commonwealth of Australia, Australian Government Department of the Environment, 2015, https://www.waterrating.gov.au/about/review-evaluation/2015-review.

Mudgal, S., 2009, *Study on Water Efficiency Standards*, European Commission (DG ENV), Paris.

The European Water Label Scheme, 2015, *Latest News from The European Water Label*, The Water Label Company Limited, Staffs, England.

Waterwise, 2017, Case for Support-supporting our vision: Water will be used wisely, every day, everywhere, UK. www.waterwise.org.uk.

Index